The Telegraph in America, 1832–1920

Johns Hopkins Studies in the History of Technology

The Telegraph in America, 1832–1920

DAVID HOCHFELDER

The Johns Hopkins University Press

Baltimore

© 2012 The Johns Hopkins University Press
All rights reserved. Published 2012
Printed in the United States of America on acid-free paper

2 4 6 8 9 7 5 3 1

The Johns Hopkins University Press
2715 North Charles Street
Baltimore, Maryland 21218-4363
www.press.jhu.edu

Library of Congress Cataloging-in-Publication Data

Hochfelder, David, 1965–
The telegraph in America, 1832–1920 / David Hochfelder.
p.   cm. — (Johns Hopkins studies in the history of technology)
Includes bibliographical references and index.
ISBN 978-1-4214-0747-0 (hdbk. : alk. paper)
ISBN 978-1-4214-0797-5 (electronic)
ISBN 1-4214-0747-7 (hdbk. : alk. paper)
ISBN 1-4214-0797-3 (electronic)
1. Telegraph—United States—History. I. Title.
TK5123.H63 2013
384.10973´09034—dc23   2012012928

A catalog record for this book is available from the British Library.

Special discounts are available for bulk purchases of this book.
For more information, please contact Special Sales at 410-516-6936 or
specialsales@press.jhu.edu.

The Johns Hopkins University Press uses environmentally friendly
book materials, including recycled text paper that is composed of at least
30 percent post-consumer waste, whenever possible.

CONTENTS

Illustrations follow page 100.

I am grateful for this pleasant duty to thank everyone who helped make this book possible. Although I have made every effort to include all who contributed tangibly and intangibly to this project, I intend no slight to anyone not mentioned here.

Many people have improved this book through their close reading and perceptive critiques. I especially thank my colleagues at the University at Albany Rick Fogarty, Carl Bon Tempo, and Susan Gauss for reading the entire manuscript and making key suggestions that improved its argumentation and readability. Merritt Roe Smith also read the entire manuscript and provided a useful critique at an important point in the revision process. I am also grateful to many others who have read portions of the book or articles and conference papers that have shaped it over the past several years. They include Kendra Smith-Howard, Richard Hamm, Brad Scharlott, Ann Fabian, Richard Bensel, Pamela Walker Laird, Paul Israel, Louis Carlat, Tom Jeffrey, Theresa Collins, Lisa Gitelman, Neil Barton, Chris Beauchamp, and Barney Finn. Richard R. John has provided key intellectual and material support over the past several years, including critiquing various incarnations of this project and graciously sharing some of his notes.

Much of the primary-source research that went into this project would have been impossible without the helpfulness and expertise of many archivists at several repositories. I especially appreciate the assistance of Steve Wheeler and his staff at the New York Stock Exchange Archives; Alison L. Oswald, Craig Orr, and Wendy Shay of the Archives Center, National Museum of American History, Smithsonian Institution; Marc Rothenberg, Frank R. Millikan, and Kathleen W. Dorman of the Smithsonian Institution Archives; Sheldon Hochheiser, formerly of the AT&T Archives, now at the IEEE History Center; Jan Hilley and Ted O'Reilly of the New-York Historical Society Library; Nicholas Noyes and

William D. Barry of the Maine Historical Society; and Melissa Mead of the Rush Rhees Library, University of Rochester.

Portions of this book have appeared in substantially modified form as articles in three journals. Some of the material in chapter 2 originated in "The Legacies of the Postal Telegraph Movements in Great Britain and the United States, 1866–1920," *Enterprise and Society* 1 (Dec. 2000), 739–61. Chapter 4 had its genesis in "'Where the Common People Could Speculate': The Ticker, Bucket Shops, and the Origins of Popular Participation in Financial Markets, 1880–1920," *Journal of American History* 93 (Sept. 2006), 335–58. Some of chapter 5 was published as "Constructing an Industrial Divide: Western Union, AT&T, and the Federal Government, 1876–1971," *Business History Review* 76 (Winter 2002), 705–32.

I also wish to thank my close friends for their support and encouragement over the years, especially Steve Hoffman, Billy Duffy, Bob Graham, and Jeff Yost. My mother, Mary Fellows, has also been a source of inspiration and love in more ways than she can imagine.

This book is dedicated to my wife, Ann Pfau. She has read every word several times, endured countless conversations about it, and lived with this project for the past fifteen years. Much more importantly, she is the source of all that is true and good and beautiful in my life.

The Telegraph in America, 1832–1920

Why the Telegraph Was Revolutionary

In January 1837, Captain Samuel C. Reid, a celebrated naval hero of the War of 1812, petitioned Congress to build a telegraph line between New York and New Orleans. Reid claimed special expertise in telegraphy, particularly that, for the past fourteen years, he had operated a line in New York harbor to announce ship arrivals. He touted his system as the ideal choice for a line of telegraphs running down the Eastern Seaboard, claiming that it was capable of sending a message from New York to New Orleans in a remarkably short two hours. Spurred on by Reid's memorial, Congress directed Secretary of the Treasury Levi Woodbury to study the "propriety of establishing a system of telegraphs for the United States," whereupon Woodbury sent a circular letter to customs collectors, commanders of revenue cutters, scientists, and other interested parties. Woodbury reported back to Congress in December and appended seventeen replies to his circular. All of Woodbury's respondents agreed that a government-owned telegraph would prove useful for national defense, official government correspondence, and commerce.[1]

Congress, Woodbury, and all but one of his respondents had in mind an optical telegraph, a straightforward mechanical technology that could have been deployed at any time in human history, particularly after the invention of the telescope in the early seventeenth century. As the name suggests, optical telegraphs used wooden arms placed on towers to send signals according to a prearranged code. The Chappe brothers set up the first working optical telegraph in 1794, in revolutionary France, and similar lines entered service in several European countries, India, Egypt, and the United States in the early nineteenth century. Nearly all of these lines were exclusively for official government use, the notable exceptions being short American lines, like Reid's, that announced ship arrivals. When Secretary Woodbury sent his circular in 1837, the federal government could well have developed optical telegraphy into a national communication network like the French system.[2]

Woodbury received one reply from an unlikely source: a painter and professor of fine arts at New York University. Little known outside of artistic circles, Samuel F. B. Morse used Woodbury's circular to bring his novel plan of long-distance communication to the nation's attention. He claimed that his own telegraph, powered by electric batteries and using electromagnets to record messages onto a moving strip of paper, could operate at night and in any weather and fit on a tabletop. Thus, he asserted, his telegraph was cheaper to build and operate than existing optical telegraphs. By harnessing the recently discovered relationship between electricity and magnetism, Morse promised that his telegraph was capable of "results of almost a marvellous character."[3]

We know the ultimate result of Morse's letter. After six years spent refining his telegraph and petitioning Congress for money to build a demonstration line, Morse in 1843 received an appropriation of $30,000 to connect Washington and Baltimore, which he did in May 1844. Morse's telegraph worked as he had claimed and soon became a commercial success. By the Civil War the telegraph network covered the country, and Morse's instrument had become synonymous with electric telegraphy in the United States and throughout the world. Morse's electromagnetic telegraph launched the electrical communication and information revolutions that continue unabated today.

Telegraphy was thus a revolutionary technology—in both senses of that term, as a revolution in technological practice and as a transformative technology with far-reaching effects on American life.[4] The telegraph broke from existing mechanical technologies precisely because it relied on electrical technology. Morse stood among the first to harness the poorly understood electrical "fluid." Although the early telegraph had roots in the world of the machine shop and networks of skilled mechanics,[5] Morse's first line in 1844 made it clear that telegraph builders needed to understand the mysteries of electricity. Unlike mechanics, electricians could not use direct tactile or visual observation to interact with telegraph equipment. Instead, they relied on indirect measurement using instruments like galvanometers. Furthermore, the telegraph was the first major technology to arise out of recent scientific discoveries in electricity. Telegraph electricians had to keep abreast of developments in battery electrochemistry, electrical circuit theory, and electromagnetism. Before the electric telegraph, few new technologies owed their origins to scientific research. After it, few did not.

Telegraphy also differed from most mechanical technologies because telegraph lines were dynamic systems. Electricians grappled with problems involving the integration of individual components into larger networks of telegraph lines and stations. The development of self-adjusting relays that automatically responded to changing line conditions, for example, became a significant technical problem as the telegraph network spread across the country. After about 1870 the industry adopted transmission methods that could transmit several messages simultaneously over a single wire. These instruments were particularly sensitive to changing line conditions. The telegraph was a mechanically simple yet electrically sophisticated technology.[6]

More importantly, the telegraph revolutionized the way people communicated and obtained information. This revolution was not instantaneous, and it affected Americans differently according to class, region, and other demographic characteristics. But it had profound effects. The line of technological development the telegraph sparked and the social changes it wrought proved as significant to the human experience as the invention of writing in the ancient world and the printing press revolution of early modern Europe. The electric telegraph forever liberated communication from transportation. All the social effects of telegraphy ultimately derive from that simple fact.

The story of the telegraph leaves no doubt that we must consider technological change and its social effects together. While cultural, business, and political considerations shaped the ways in which telegraphy transformed American life, change arose from—and did not exceed—the specific attributes of the technology itself. Furthermore, individuals and institutions did not simply adopt the telegraph—they adapted themselves to it. Telegraphy did not fulfill its potential as a driver of social change until its users reshaped their actions, organizations, perceptions, and expectations around it.[7] This development occurred fully after about 1860.[8] When users gradually stopped shaping themselves according to the telegraph, after about 1920, telegraphy's importance correspondingly diminished.

Thus the telegraph enjoyed its heyday roughly between 1860 and 1920, when it most strongly affected American life. It came of age during the Civil War as battlefield commanders and military and civilian leaders learned how to exploit it to coordinate operations and strategy. Its ruggedness and mechanical simplicity allowed military telegraphers to take down and set up their equipment rapidly, thus aiding field commanders to direct logistics and operations. More importantly, the question of who ought to control the military telegraph system became a major point of conflict between civilian and military officials.

The Civil War also accelerated an existing trend toward consolidation in the telegraph industry, partly because the economics of the industry encouraged concentration and partly because the war struck the major telegraph companies differently. In 1866 Western Union bought out its last rivals; thereafter it remained the nation's dominant telegraph company, handling at least 80 percent of telegraph traffic in the country. During the last third of the nineteenth century, Western Union's managers, promoters of rival telegraph companies, state and federal legislators, political economists, and antimonopoly reformers all grappled with the consequences of this new form of business enterprise—the so-called natural monopoly. Western Union's dominance of the telegraph industry brought about, indeed forced, the working out of important elements of public-utility economics and policy.

During the late nineteenth century, many antimonopolists sought to place telegraphy under federal regulation or operation. They feared Western Union's control of the industry mainly because the nation's news and financial information flowed through the company's wires and thus were absolutely dependent upon them. Newsbrokers had relied on the telegraph to produce and distribute the news since the mid-1840s, and reporters and editors quickly discovered that the economics of telegraphy enforced brevity and encouraged cooperative newsgathering. By the Civil War it had become apparent that the telegraph was transforming the production and distribution of news. At the same time, wire-service journalism changed how Americans consumed the news, instilling modern expectations about timeliness and newsworthiness. Many contemporaries also expected the press to be the main vehicle for the telegraph to act upon written language more generally, that it would streamline and invigorate American literary style. But the telegraph had a muted effect upon literary style because the high cost of telegrams restricted its use to a small minority of the population.

Just as the telegraph changed the business of newsgathering, so did it reshape the nation's financial markets. The stock ticker, invented in 1867, allowed brokers to monitor markets at a distance from exchange floors. Within a few years, the ticker changed exchange operations, recast relations between brokers and customers, and transformed the overall structure of the nation's financial markets. The ticker also reconfigured the psychology and geography of financial markets. Traders came to regard markets less as places to exchange tangible goods and more as the flow of ticker quotations posted on blackboards in distant offices. As financial markets increasingly became markets in information, the telegraph network laid the technological foundation for dispersed stock trading and ownership in the twentieth century, a hallmark of modern capitalism.

By 1900 or so, the telephone began to supplant the telegraph as the premier electrical communication medium. The intertwined history of the two industries helps us to understand the process of industrial and technological succession. Although the telephone originated in efforts to increase the message-carrying capacity of telegraph lines, the long-distance telephone by about 1890 had begun to cut into telegraph revenues. In 1909 American Telephone and Telegraph (AT&T) acquired control of Western Union—signaling the telephone's eclipse of the telegraph. Although AT&T disgorged Western Union in 1913 under threat of antitrust prosecution, the telegraph industry had devolved to junior-player status in the nation's long-distance communications market. Even so, had Western Union marketed its services more aggressively and developed an internal research-and-development capability, the telegraph industry's trajectory in the new century might have been quite different. Instead, the telegraph industry's survival increasingly depended on the willingness of the Federal Communications Commission to prop it up as the only meaningful competitor to AT&T's telephone empire. With the deregulation movement of the 1970s and 1980s and the increasing adoption of fax and email, telegraphy lost both its regulatory protection and its rationale as a record-communication medium. The ultimate fate of the telegraph depended as much on social preferences and political goals as it did on economic and technical considerations.

CHAPTER ONE

"Here the Telegraph Came Forceably into Play"

THE TELEGRAPH DURING THE CIVIL WAR

In August 1862, William L. Gross, a twenty-three-year-old lawyer just admitted to the Illinois bar, decided to volunteer for the Union army. Explaining that "my conscience smote me and I could stay at home no longer," Gross raised a company of volunteers in and around Dwight, Illinois. However, on the day that Gross telegraphed Governor Richard Yates that his company was ready for duty, John Van Duzer, an official of the U.S. Military Telegraph Corps (USMT), offered him a post as manager of the Cairo, Illinois, military telegraph office for sixty dollars a month. Although Gross admitted that he was "an indifferent operator" who had not touched a telegraph key in six months, the USMT was in short supply of experienced telegraphers to staff its large and expanding network. Gross promptly accepted Van Duzer's offer, noting sardonically in his diary, "Thus ended my brief military career." When he mustered out in 1866, however, Gross held the rank of lieutenant colonel and assistant superintendent of the USMT. During the war Gross became a protégé of Norvin Green, president of the South Western Telegraph Company. When Green became vice-president of Western Union after the great telegraph mergers of 1866, he brought Gross with him to the company's New York headquarters. After a bitter feud with Western Union superintendent Anson Stager over the company's rate structure, he resigned as chief of the Tariff Bureau in early 1868, returning to his Illinois law practice.[1]

Gross's experiences help us understand three aspects of the history of telegraphy during and immediately after the Civil War. Most importantly, the telegraph proved its value as a tactical, operational, and strategic communications medium. Perhaps its most significant function in this regard was to help civilian officials maintain control over military operations. For telegraphers, shared wartime experiences helped forge a craft identity, manifested both in the pride they took in their skills and in their incipient trade unionism. During the war, demand for telegraphers was high, both in the USMT and on company lines, and operators enjoyed good wages and considerable autonomy. For many telegraphers who remained in the industry after the war, their service in the USMT was a counterpoint to the wage reductions and degradations they experienced under Western Union management. The major reason why Western Union was able to impose greater control over its postwar work force was that the war accelerated an ongoing trend toward consolidation in the telegraph industry. Largely as a result of wartime developments, in the summer of 1866 Western Union bought out its two major rivals, establishing it as the dominant force in the nation's telegraph industry thereafter.

The Telegraph as a Military Tool

From 1 May 1861 to 30 June 1865, the USMT handled some 6.5 million messages at a total cost (for construction, repair, and operation of the network) of $2,655,000, or about $0.41 a message. During the war the USMT built fifteen thousand miles of line, often in adverse conditions and sometimes under enemy fire. At its peak in 1865, the USMT network consisted of more than eight thousand miles of military telegraph line and another five thousand miles of commercial lines in the South operated by military telegraphers.[2] The USMT amply repaid this investment. The network repeatedly proved its value as a strategic and logistical tool in the field. Indeed, the USMT was one of the keys to a northern victory. Confederate officials were never able to exploit fully the southern telegraph network for military purposes.[3] Just as importantly, the military telegraph enabled political leaders to maintain civilian authority over military operations and to control the flow of news. Secretary of War Edwin Stanton referred to the military telegraph as his "right arm," and Secretary of State William H. Seward received advance summaries of European steamer news through the USMT. President Lincoln, as is well known, spent countless hours in the War Department telegraph office adjoining Stanton's office.[4]

During the winter and spring of 1861, however, northern leaders had yet to recognize the military value of the telegraph. The need for some sort of control over the telegraph had become clear in the last weeks of Buchanan's administration, before actual fighting had begun. As early as January 1861, some southerners and their northern sympathizers ordered weapons and relayed sensitive information about government decisions through the telegraph. To stop such messages, officials of the American Telegraph Company and other lines began to censor suspect telegrams informally, blocking those that seemed to aid the growing secession movement. From April to October 1861, the American Telegraph

Company, whose lines ran the length of the Eastern Seaboard from Nova Scotia to New Orleans, provided the bulk of telegraph facilities for the Union government and military.[5]

The war's first casualties were men of the Sixth Massachusetts Regiment sent to defend Washington in April 1861. On 19 April a secessionist mob attacked them while they changed trains in Baltimore, leaving four soldiers and nine civilians dead. Two days later, Maryland state officials cut the wires leading north from Washington to Baltimore, leaving the capital without telegraphic communications northward. At the same time, the Baltimore and Ohio Railroad Company refused to transport government troops and military supplies. Cut off from the rest of the country and with only two thousand troops to defend the city, government officials and residents feared an imminent Confederate occupation of the capital.[6]

Washington's dangerous exposure impressed on Secretary of War Simon Cameron the need for military oversight of both the telegraph and railroad networks. He asked Thomas A. Scott, general manager of the Pennsylvania Railroad, to organize the military railroads and telegraphs. Scott gave the task of organizing a military telegraph network to his protégé Andrew Carnegie, who asked in turn his boyhood friend David McCargo, superintendent of the railroad's telegraph line, to send his four best operators for service at the War Department. These four operators reported to the War Department at the end of April. Their first task was to connect the various military sites in and around the capital to the War Department. Because Congress had yet to appropriate funds for a military telegraph, the American Telegraph Company paid for the construction and operation of these lines.[7]

Meanwhile, in the West, George McClellan was an early adopter of the telegraph for military purposes. In April 1861 Ohio's governor William Dennison agreed with McClellan that military supervision of the state's telegraph lines was necessary "to stop all messages of a disloyal character," especially "orders for munitions and provisions for the South." McClellan tapped Western Union superintendent Anson Stager, based in Cleveland, for this task. He asked for similar authority from the governors of Illinois and Indiana, explaining to Governor Yates of Illinois that since "the telegraph is not a military organization," Stager needed the governor's consent to assume this control. Armed with the governors' permissions, at the end of May McClellan appointed Stager "superintendent for military purposes of all the telegraphic lines within the Department of the Ohio." McClellan's order and the governors' grant of authority indicated the hybrid nature of the military telegraph system. Indeed, William R. Plum, military telegrapher and later author of a two-volume history of the service, remarked that at the time "military control of private lines was merely nominal."[8]

By October the ad hoc military telegraphs already consisted of about three hundred miles of newly constructed lines, about fifty stations, and eighty-three operators. At this time, Stager recognized the growing importance of the telegraph for strategic military communications and presented a plan to Cameron outlining a comprehensive military telegraph system. While Stager envisioned a light military supervision, Assistant Secretary of War Thomas Scott added the provision that the War Department could take possession of any telegraph lines if military necessity required it. At the end of October, Lincoln and Cameron agreed to adopt Stager's plan. Though Stager did not consider it necessary to give military ranks to telegraphers, Quartermaster General Montgomery Meigs insisted on it in order to streamline ordering of telegraph supplies and equipment. Cameron gave Stager the rank of captain and assistant quartermaster and then a promotion to colonel in February 1862.[9]

On 1 November 1861, President Lincoln gave General McClellan command of all Union armies as general-in-chief, replacing the aged Winfield Scott. One of his first acts was to order generals in the field to build telegraph lines connecting their headquarters with his.[10] By January 1862, the country's two major telegraph companies, the American and Western Union, ran these lines into McClellan's Washington, D.C., headquarters. A few weeks later, however, Secretary of War Edwin Stanton (who had replaced Cameron in late January) ordered the lines removed from McClellan's headquarters and connected them instead into the War Department offices. For the remainder of the war, Stanton controlled the army's communications and (along with Assistant Secretary of State Frederick Seward) oversaw the censoring of telegraphic news dispatches.

The military telegraph network proved its value in coordinating broad strategy almost immediately and continued to do so throughout the war. As McClellan advanced from Ohio into northwestern Virginia in the summer of 1861, he built a line to keep in touch with his subordinate commanders. By the end of the summer, McClellan's military authority extended to include the volatile border state of Missouri, and the telegraph allowed him to oversee events there while he operated in the Kanawha Valley some six hundred miles away. On 16 February 1862, just hours after the fall of Fort Donelson, McClellan engaged in a three-way real-time conversation with Henry Halleck and Don Carlos Buell to discuss plans for advancing to Nashville. Similarly, Ulysses S. Grant later recalled that he had "held frequent conversations over the wires" about strategy with Stanton during 1863, some lasting two hours. William Tecumseh Sherman also recalled the "perfect concert of action" between his forces in Georgia and Grant's in Virginia in 1864. "Hardly a day intervened when General Grant did not know the exact state of facts with me, more than fifteen hundred miles off, as the wires ran."[11]

The military telegraph also proved valuable on several occasions as an operational and tactical tool on the battlefield, allowing commanders to remain in constant touch with subordinates and to react quickly to changing conditions. McClellan adroitly used the telegraph to resupply his troops with bullets and shells in the midst of the Battle of Antietam, Maryland, in September 1862. Assistant Secretary of War Charles A. Dana later praised the utility of the telegraph when he accompanied Union forces during the Battle of Chickamauga in northern Georgia on 19 September 1863, noting that "it was one of the most useful accessories of our army," giving General Rosecrans "constant information on the way the battle was going." Also, Dana was also able to send eleven telegrams to Washington, apprising Stanton of the progress of the battle on almost an hourly basis.[12]

At the Battle of Spotsylvania during the Wilderness campaign of May 1864, General George Meade used the telegraph to reinforce General Winfield Scott Hancock's Second Corps after it had come under heavy Confederate counterattack. Luther Rose, a USMT telegrapher attached to Hancock's headquarters, set up his key and sounder at 3:30 AM, an hour before Hancock's advance on the Confederate lines, allowing Hancock's chief of staff to coordinate the attack with other corps commanders. Favored with a heavy early morning fog, Hancock's advance was successful. Later in the day, however, the Confederates counterattacked. Hancock telegraphed to Meade that he was unable to hold his gains unless the Sixth Corps on his right came to his support. Ten minutes later, as Rose recorded in his diary, "the 6th Corps was thundering away & Hancock held his own. . . . Here the Telegraph came forceably into play, showing to what great benefit it could be used." Rose used a field telegraph consisting of light iron wire insulated with India rubber. It could be deployed within a few minutes from the backs of mules and could be strung almost anywhere. Such flexibility meant that Rose accompanied Hancock closely, taking down and resetting his instruments if Hancock moved his headquarters more than half a mile. Rose and a companion operator were so close to the front at Spotsylvania that heavy shelling frequently broke their wire. The two took turns splicing the breaks, remarking before setting out, "If I stop a shell, send my things home."[13]

Rose later described his telegraph instrument as "the principal channel" through which passed the orders determining the movements of Hancock's corps during the Wilderness campaign. Similarly, Meade later recalled that during the ill-fated Battle of the Crater at Petersburg on 30 July 1864 he had sent or received more than one hundred telegrams during the five-hour battle, at the rate of one every three minutes. Rose himself operated from an artillery battery during that engagement, demonstrating the utility of the telegraph for real-time battlefield use.[14]

Despite the usefulness of the telegraph as a strategic and tactical communication medium, the USMT had one important limitation—it remained a civilian organization. The result was an uneasy hybrid, a telegraph system that served the military but was not part of it. While the USMT built thousands of miles of its own lines to link military commanders with each other and the War Department, it relied heavily on the existing commercial telegraph network. Telegraph companies gave priority to military and government messages, while continuing to handle commercial traffic and earn enormous wartime profits. The top dozen officials of the USMT were all officers in the Quartermaster Corps, yet they retained their civilian positions as managers of the commercial lines. Many of the approximately twelve hundred operators and linemen in the USMT also continued to work for commercial telegraph companies and drew only part of their salaries from the War Department.

The USMT held this ambiguous status partly because the Civil War was the first war in which military leaders relied on electrical communication. Although the Mexican and Crimean wars had demonstrated the powerful newsgathering capabilities of the telegraph, neither war saw the medium used extensively for military purposes. Northern military and political leaders had to learn how to use the telegraph as the war unfolded. Furthermore, the United States Army had only recently come to recognize the importance of military signaling in general. Colonel Albert J. Myer (himself a former telegrapher) had been appointed as the first chief signal officer of the army only in June 1860. Myer had no trained telegraphers in the Signal Corps when the war broke out. Perhaps the most important reason why the USMT remained a civilian organization was that key members of the Lincoln administration, particularly Stanton, wished it. Stanton regarded his supervision of the telegraphs as central to maintaining control over both military operations and the flow of news.

The USMT's independence from the military chain of command led to two important problems for military operations. To begin with, most USMT personnel were not subject to military authority, a frequent source of irritation to commanders in the field. Although USMT operators usually performed their duties without hesitation, they often refused to conform to military standards of discipline. More seriously, the USMT rarely coordinated its operations with those of the Signal Corps, resulting in a lack of integration of electric telegraphy with other forms of signaling. Myer continually lobbied to have battlefield telegraphing, if not all of military telegraphing, placed under his purview as part of the Signal Corps. Myer's continuing opposition to Stanton's wishes ultimately resulted in his removal as chief signal officer in November 1863.

The USMT drew high praise from many field commanders in their wartime reports and later memoirs. After the war, former USMT operators seeking military pensions and admittance to the Grand Army of the Republic obtained testimonials from a dozen generals attesting to the military value of their service.[15] Yet civilian control of the USMT arguably fostered a lack of cooperation and even insubordination among telegraphers. Moreover, many USMT operators who were reluctant to share hardships at the front deserted their posts for the safety of the rear. Not surprisingly, contemporary sources and postwar commentators described some USMT operators as "venal," "cowardly," "lazy," and "drunken" while on duty.[16]

The wartime correspondence and diaries of several USMT operators paint a more favorable picture. Men like William Gross and Luther Rose performed their duties to the best of their abilities. Of the approximately ninety-five thousand court-martial cases during the Civil War, only six involved telegraph operators. Two of the cases arose because operators had been given conflicting orders by different generals. Only one ended in a conviction, but it resulted in a commuted sentence with the operator immediately returning to duty. Had USMT operators been as insubordinate and useless as some suggest, more telegraphers would have been hauled before courts-martial. Furthermore, the leaders of the USMT, Superintendent Anson Stager and the several assistant superintendents, directed telegraphers under their supervision to obey the orders of commanding officers at their posts.[17]

However, USMT telegraphers often did engage in conduct that no military officer would tolerate from subordinates. Many of the operators under Gross's supervision, for instance, drank on duty, failed to answer morning roll calls, and took bribes to transmit messages of cotton speculators ahead of military traffic. On at least two occasions, USMT telegraphers threatened strikes if their demands for better pay and working conditions went unmet. The operators backed down after General Grant and Colonel Stager threatened to arrest them.[18]

On their part, officers often expected the USMT to provide instant communications, even when operators were unable to do so. In February 1862 General Halleck tried to have J. J. S. Wilson (assistant superintendent of both the USMT and Illinois and Mississippi Telegraph Company) fired for delaying military traffic at Chicago while permitting commercial messages to pass freely. Stager intervened and told Halleck that the reason for the delay was that flooding had broken a cable crossing the Ohio River at Paducah, Kentucky. Halleck brusquely replied, "Remedy the defect. . . . There must be an end to this inefficiency and delay." In 1864 Theodore Holt, an operator assigned to General Eugene Carr at Little Rock, Arkansas, could not raise a nearby office, so Carr forced Holt to operate under armed guard. On another occasion a major general threatened to shoot Gross if he did not send a certain message to a distant office within the hour. USMT operators frequently complained in their diaries and letters of such treatment, expressing sentiments as Holt did, that "Carr don't own the Telegraph Corps."[19]

More seriously, the USMT did not coordinate its operations with the army's Signal Corps. In many ways, the USMT was on firmer footing than the Signal Corps for the first two years of the war. At the beginning of the war, Myer faced problems in staffing the new corps. Without a formal organization of its own, the corps relied on officers and enlisted men detailed from their home units, who could be recalled suddenly at the will of their commanders and were not eligible for promotion while on detail. Thus, many personnel avoided signal duty as harmful to their careers. In his first two annual reports to the secretary of war in November 1861 and 1862, Myer strongly pressed for the permanent establishment of the Signal Corps. Congress passed legislation establishing the corps for the duration of the war in February 1863. Afterward, the corps came to comprise just under eighteen hundred officers and enlisted men, with about another seven hundred detached to it for temporary duty.[20]

To remedy the lack of coordination between the electric telegraph and other forms of military signaling, Myer repeatedly sought to bring telegraphy within his duties. Because he regarded signaling as necessary to tactics and operations, he believed that all personnel engaged in military signaling, whether by signal flags or electric telegraphs, should have been subject to military discipline. In June 1861, for instance, he asserted to Secretary of War Cameron that his commission gave him "general charge of the telegraphic duty of the Army, whether . . . transmitted by means of Electricity or by aerial Signals." In July he asked his friend John Callan, clerk of the Senate Committee on Military Affairs, to intercede for him with that committee to get control over the telegraph, since by that time he had already regarded the USMT as a dangerous rival to his Signal Corps. In August he again asked Cameron to allow the Signal Corps to recruit trained telegraphers "who shall be instructed also in the use of the telescope and aerial signals, and who, employed for the war, shall be sworn to the faithful discharge of their duties." Myer also asked for $30,000 to develop his field telegraphs, a sum that Congress refused to appropriate despite hearty endorsements from Generals McDowell and McClellan. Finally, on 14 August, Assistant Secretary of War

Thomas Scott authorized Myer to develop a field telegraph out of money earmarked for the USMT.[21]

Myer envisioned this field telegraph to be a rugged and simple instrument that could be operated by soldiers unfamiliar with Morse code and the mysteries of electricity. Supplied with Scott's authorization, he contracted with Henry J. Rogers of Baltimore to build such a telegraph. Rogers was an ideal choice. An experienced telegrapher who had been one of Morse's first operators on the pioneer Washington-to-Baltimore line in 1844, he had also developed a system of maritime signals for merchant vessels that he modified for the navy during the war.[22] He was thus familiar with both flag and electric signaling and how they might function together in the field. In January 1862 Rogers tested his field telegraph at the Signal Corps's Georgetown training camp. His telegraph consisted of a sealed battery that required less maintenance than contemporary wet cells and a dial faceplate showing all the alphabet letters and numerals. By the spring of 1862, New York electrical manufacturer George Beardslee had modified Rogers's original design by replacing the battery with a hand-crank magneto. The field telegraph trains carried reels of insulated wire that could be rapidly laid on the ground or on short lances.

In theory, the Beardslee instrument was well suited for battlefield use. But it provided mixed results in actual operation and soon became a source of conflict between Myer's Signal Corps and Stager's USMT, revealing their vastly different views of military communication. The Beardslee field telegraph was designed for rapid deployment and ease of use by troops lightly trained in electric telegraphy but who also knew other forms of signaling. In contrast, the USMT relied on relatively fixed telegraph lines manned by civilian operators who were experienced in reading Morse code by sound and understanding electrical phenomena but did not know military signaling methods.

Four of the Beardslee telegraphs accompanied McClellan's Army of the Potomac during the Peninsula campaign of early 1862. The corps relied heavily on visual signaling during that campaign, using the field telegraph only to connect McClellan's headquarters to the rather static front. McClellan continued to rely on the USMT first, regarding the Signal Corps as an auxiliary to it. He nevertheless spoke well of the field telegraph's usefulness in reaching areas not connected by the USMT's lines. The Beardslee system also had some success at Fredericksburg in December 1862, when smoke so obscured the battlefield that visual signaling could not be used. Its success here allowed Myer to obtain more funding for additional telegraph trains, and by late 1863 he had placed about thirty in service. Despite limited battlefield successes and two years of development work, the field telegraphs never performed properly. They were prone to damage by lightning and were not rugged enough to stand service in the field. The Battle of Chancellorsville in April and May 1863 exposed the shortcomings of the Beardslee instrument, leading Signal Corps operators to turn over some of its lines to USMT operators, who used pocket Morse instruments instead. Afterward, Myer converted some of his field telegraph squads from the Beardslee instrument to Morse keys and sounders.[23]

Seen from a purely military perspective, the lack of coordination between the Signal Corps and USMT was unfortunate. Myer was correct in theory that the field telegraphs were properly the purview of the Signal Corps. From a wider perspective, however, an independent USMT gave civilian government officials greater control over military affairs. Secretary of War Stanton especially placed great importance on controlling the telegraph network outside of the military chain of command. War Department clerk Charles F. Benjamin later gave some insight into the effectiveness of Stanton's control. He recalled that Stanton's telegrams "were so frequent, peremptory, and regardless of hours" that one general "never lay down in his tent or quarters without a mental picture" of Stanton "watching his every movement." An early biographer of Stanton called his control of the telegraph "a perfect autocracy" and the USMT "a part of his own personal and confidential staff." Critics, including General George McClellan and Secretary of the Navy Gideon Welles, charged that Stanton wanted to control the telegraph mainly to increase his influence within the Lincoln administration. However, his close supervision of the telegraph firmly placed generals under civilian control at a time when military influence over the country's affairs was at a peak. McClellan, for example, speculated in the summer of 1861 that he could "become Dictator" if he so chose.[24]

Despite Stanton's well-known intention to maintain civilian control of the telegraph, Myer continued to press for Signal Corps operation of the field telegraphs. Myer's methods verged on insubordination because of his strong belief that the private telegraph companies were bent on throttling his corps. In particular, Myer attributed Stanton's resistance to the influence of senior telegraph managers such as Anson Stager, Thomas Eckert, and Edward Sanford. In April 1862, for example, Myer sent a circular to his signal officers, claiming that the corps's "permanent existence" was threatened by "interested and designing men outside of the Corps"—by which he meant the managers of the USMT—and asked them to obtain letters of support from the generals they served so that he could lobby Congress with them. In the wake of the abandonment of the Beardslee telegraphs after Chancellorsville, Myer placed an advertisement in an army periodical for experienced telegraph operators. Though he claimed that he needed only thirty to sixty operators, this advertisement earned him a reprimand from Stanton. Stanton's rebuke led him to issue another circular to his officers, charging that the American Telegraph Company was attempting to acquire control over the field telegraphs. USMT superintendent Anson Stager regarded Myer's advertisement for operators as an encroachment on his efforts to recruit and retain the same class of operators, leading Stager in October to petition Stanton for control of the Signal Corps's telegraphs. Stanton backed Stager, dismissed Myer as chief signal officer in November, and turned over all field telegraphs to USMT operators, who promptly replaced the balky Beardslee instruments with pocket Morse instruments. Myer had so alienated Stanton that he briefly considered promoting Western Union's Thomas Eckert, manager of the War Department telegraph office, from major to colonel and installing him as head of the Signal Corps, a move that surely would have rankled Myer. In a private letter written just after his 1863 dismissal, Myer charged that the "continued success" of the field telegraphs had generated "a hostile feeling . . . so great that the gigantic arm of monopoly had been stretched out in greed to clutch this condemned instrumentality."[25]

After his dismissal, Myer continued to lobby both for his reinstatement and for electric telegraphy to be recognized as part of the Signal Corps's sphere of responsibility. He succeeded after the war and returned to active duty in October 1867. He remained chief signal officer until his death in 1880. Despite his fractious wartime relationship with Eckert and Stanton, he was on good terms with Anson Stager, who gave Myer technical advice after the war and recommended instructors for telegraph training at Fort Whipple. Immediately after his postwar reinstatement, Myer sought to set up a Signal Corps capable of providing military communication in a future war, promising that "if another war comes, we will try to show what the Corps with the light lines and good telegraphers can do."[26] However, the major mission of the Signal Corps from 1870 to 1890 was weather prediction. Myer's most notable postwar accomplishment was the establishment of a national weather reporting system staffed by volunteer observers and Signal Corps personnel. Despite his good working relationship with Stager, Myer had frequent conflicts with Western Union over the rates and terms of service for the weather reporting system.

Myer's contributions to military communications are hard to overstate.[27] Myer was correct that all forms of tactical and battlefield communications were properly the role of the Signal Corps. Although the Signal Corps was of limited use during the war, Myer modernized and regularized the service by setting up training camps and by publishing a *Manual of Signals*, which appeared in 1864.

In subsequent wars, whether against Native peoples in the West or in Cuba and the Philippines at the turn of the century, the Signal Corps performed all army communications functions. No hybrid organization like the USMT would exist in future conflicts. Myer's weather reporting system, taken over by the Department of Agriculture in 1891, led to fundamental strides in the new science of meteorology and prevented many shipwrecks on the Great Lakes. Yet his tenacity in attempting to acquire control of the military telegraphs and his penchant for infighting nearly ended his military career in disgrace in 1863. One officer later recalled that when he had asked General Philip Sheridan to support Myer's reinstatement as a personal favor, Sheridan replied, "All right, Howard, I'll do it. I would have done it before if Myer had not been such a d[amne]d old wire-puller."[28]

The Wartime Experience of Military Telegraphers

On 12 August 1862, William Gross arrived at the Cairo, Illinois, USMT office to begin his career as a military telegrapher. Located at the junction of the Mississippi and Ohio Rivers, the Cairo office was an important meeting point of the lines of two telegraph companies (the Illinois and Mississippi and the South Western) and was a jumping-off point for Union forces moving westward into Missouri and southward into Kentucky and Tennessee. Despite Cairo's important location, Gross was immediately dismayed at his surroundings, calling the town a "God-forsaken place" of mud and pestilence. During the war, at least eighteen USMT operators fell ill while stationed there. The USMT office was on the third floor of a building housing a saloon and billiards hall on the first two floors. The office itself was a dark and dirty room that also doubled as storage space and sleeping quarters for the operators. Gross received sixty dollars a month, soon raised to seventy, to manage the Cairo USMT office. Shortly after his arrival, he decided to spend five dollars a week of his salary for room and board at a local hotel, sharing a room with Van Duzer. Gross's duties at Cairo were relatively light, consisting of conducting morning roll call on two telegraph lines running into the office, testing the line and ordering repairs, and writing up vouchers and reports. He remained at Cairo until January 1863 when, much to his relief, he was ordered to the Memphis, Tennessee, USMT office.[29]

Gross's experiences at Cairo were typical of rear-echelon USMT personnel. His pay was within the range for operators who could copy Morse code by sound. His responsibilities were a blend of military and civilian duties: he handled both military and commercial traffic and received a supplemental salary from the South Western Telegraph Company. Similarly, John Lonergan, cipher operator for General Thomas in 1865, operated from the Nashville and Louisville Western Union offices. Because military traffic was light, he mainly handled commercial business. Gross's working conditions were also typical of both military and commercial operators. Thomas Edison, who began his career as a seventeen-year-old telegraph operator in 1864, recalled that the Louisville, Kentucky, USMT office was "positively bleak" and "dingy and likewise dirty," reeking of acid fumes from the battery room below the operating room.[30]

For many of the twelve hundred operators, constructors, and repairers who staffed the USMT, their wartime service was a formative experience. Several wrote postwar memoirs describing their service or left behind diaries and collections of letters. Others compiled histories of the USMT through personal interviews or letter correspondence with former comrades. Many helped to organize a veterans' organization, the Society of the USMT, which held its first annual meeting sixteen years after Appomattox. Their memoirs, histories, and society were part of a general trend toward postwar nostalgia as Civil War participants grew middle-aged, similar to the decline and rejuvenation of the Grand Army of the Republic.[31] Several USMT personnel would play key roles in the postwar telegraph industry. Unfortunately, southern telegraphers left little in the way of memoirs or personal papers, and they did not form an organization like the Society of the USMT. Thus, it is difficult to include the wartime experiences of southern operators.

Given the number of USMT personnel, it is impossible to make a comprehensive generalization about why they served. Many at the upper echelon of the USMT, such as Anson Stager and Thomas Eckert, were already senior managers for the commercial lines. They smoothly incorporated their new military duties into their existing responsibilities. Others, including William Gross and John Van Duzer, had been commercial telegraphers who also had managerial or professional experience that fitted them for middle-management positions in the USMT. Some, like Pennsylvanian Joseph Schnell, took to telegraphing to avoid military service while still helping the war effort. Schnell had initially served as a three-month infantry volunteer at the outbreak of the war. Luther Rose served two stints in the USMT and seemed to have had a mixture of motives. He patriotically joined the government telegraph service in May 1861, immediately after the War Department called for experienced operators. He quit after a long bout with typhoid fever but rejoined in 1863 because he missed the adventure and camaraderie. Young Thomas Edison joined the USMT to escape small-town Michigan and to see the wider world. Most of these men remained active in the telegraph industry after the war. Stager served as Western Union's general superintendent officer until 1881. Eckert served as general manager of the Atlantic and Pacific

Telegraph Company as well as Western Union, eventually becoming president of Western Union in the 1890s. Gross remained with Western Union until early 1868. Van Duzer became a telegraph instructor for the Signal Corps, ascending to the rank of colonel before leaving the service in 1877. Schnell managed the Western Union office in Binghamton, N.Y., until 1875. Edison, of course, remained central to telegraph innovation for many years.[32]

Whatever their different motivations for joining or their varied wartime postings, USMT personnel shared common experiences. Operators and linemen in rear areas, like William Gross, performed jobs not much different from civilian telegraphers and managers. Those who served in forward areas, like Joseph Schnell and Luther Rose, experienced many of the hardships and dangers of frontline troops. They all complained of poor pay, complaints that grew more strident in the face of wartime inflation. They also asserted their independence from the army officers they served, often refusing to conform to military discipline and frequently complaining that they endured second-class status in camp and at headquarters. Many were young men experiencing their first taste of freedom from parents and hometowns. Their letters, diaries, and later reminiscences were replete with stories of hard drinking, womanizing, and pranksmanship, all while performing legendary feats of telegraphing. Indeed, L. H. Smith, the editor of the *Telegrapher*, a journal that advocated for telegraphers and ran technical articles, lamented in November 1864 that "an operator is described as 'a gay fellow,' or spoken of as 'fast,' as if the more evil he plunges into, the higher the estimate put upon him."[33]

As Smith's remarks suggest, the young men who staffed the USMT could indulge in such dissipations because skilled operators were in short supply. Managers and army officers were forced to tolerate any antics short of open disloyalty to the northern cause or chronic inattention to duty. By all accounts, telegraphers—both commercial and military—were hard drinkers. George Kennan told his family that he and his fellow operators at Western Union's Columbus, Ohio, office kept a communal whiskey bottle on hand throughout the day and that major customers frequently treated them to oyster and champagne suppers. Several Federal and Confederate operators claimed to have tricked their post commanders into supplying them with whiskey, explaining that it could substitute for battery acid, but only if it was of the finest quality. In April 1864 USMT assistant superintendent Robert Clowry directed the chief operator at Little Rock, Arkansas, to ensure that drinking on duty be "confined to the bounds of moderation," because he could not stop it outright. In June 1865 Clowry fired an operator for thievery, complaining that discharged operators "make their boasts that they can readily get employment from com'l co's." Similarly, J. J. S. Wilson, superintendent of the Illinois and Mississippi Telegraph Company, was forced to promote—instead of firing—a skilled yet dissipated young operator, lamenting that "if operators were plenty I would not risk keeping him anymore."[34]

Another indication of the high demand for skilled telegraphers was their exemption from military service. The War Department exempted them in August 1862. In 1863 the American Telegraph Company offered to pay the $300 exemption fee for any of its operators who were drafted, provided they agreed to stay with the company for a year at their present salaries. The commercial lines, earning fabulous wartime profits, could afford to pay far better than the USMT did. To make matters worse, USMT telegraphers often faced delays receiving their pay, in some cases up to four months in arrears.[35]

Telegraphers resorted to several strategies to cope with their low pay and harsh working conditions. Operators who staffed offices in rear areas often did double duty for the commercial lines and received supplemental pay from them. In addition to supervising about 125 USMT personnel, William Gross managed twenty offices for the South Western Telegraph Company for another fifty dollars a month. A. T. Langhorne, a repairer working for Gross, received forty dollars from the USMT and thirty-five from the South Western Telegraph Company. Some, like Clowry's operators in Arkansas, refused to turn over office receipts until they received their back pay. Both northern and southern operators took bribes from cotton and stock speculators to give their messages priority over military and other commercial traffic, or engaged in speculation themselves.[36]

Telegraphers both North and South also organized to improve their conditions and assert their autonomy. Indeed, the Civil War marked the birth of trade unionism among telegraphers. In November 1862 Grant tried to remove Van Duzer, the USMT manager in charge of his area of operations, and to replace him with his own aide Colonel John Riggin. In response, the operators working for Van Duzer nearly walked off the job. Only Grant's threat to arrest them and Van Duzer's reinstatement prevented a walkout. A month later, fifty-six USMT operators signed a petition complaining that they were paid less than comparable civilian employees of the Quartermaster and Commissary departments. Their petition, forwarded to Secretary of War Stanton, had no effect. A more serious organizing effort, almost resulting in the only strike among northern telegraphers, occurred in William Gross's division in Kentucky and Tennessee in January 1864. Nineteen operators signed a petition demanding more pay and threatening to resign in a body if their demand went unmet. Gross forwarded this petition to Stager, who responded that he regarded this petition as a treasonous conspiracy and threatened to arrest them if they walked out. Gross smoothed matters over, and the operators returned to their keys without further incident. In the fall of 1863, northern telegraphers organized the National Telegraphic Union (NTU), partly a trade union and partly a social and educational organization that published the *Telegrapher*. The craft solidarity that emerged during the Civil War would energize the telegraph strikes of 1870 and 1883.[37]

Southern operators faced even worse pay than northerners did because of much higher inflation in the Confederacy. As a result, southern operators were far more militant. In the fall of 1863 they formed the Southern Telegraph Association (STA), much more a trade union than the NTU, its northern counterpart. They asserted that the Southern Telegraph Company, the major commercial line in the South, forced them to work long hours, often fifteen hours a day, at poor wages. That company's president, Dr. William Morris, responded by threatening to fire all operators belonging to the STA and to forward their names to local conscription officers. The STA retaliated by forbidding its members from handling any traffic save official government or military messages. Twenty-two operators elected to obey the STA's order, resulting in a strike that lasted a week. True to his word, Morris broke the strike by discharging the strikers and subjecting them to conscription; most of them returned to work chastened. This episode was notable for being the Confederacy's only labor strike and the first undertaken by telegraphers.[38]

Besides low pay, the most common complaint among USMT operators was their lack of status and respect, mainly because of the hybrid nature of the USMT itself. Operators in camp frequently grumbled that they were treated like enlisted men, not befitting practitioners of a skilled and arcane craft. Assistant Superintendent John Van Duzer complained at length to his superior Samuel Bruch that morale was low among his operators because "they have no recognized status, and every officer, of whatever grade, who happens to command a post where an office is maintained, has to be taught that the operator is not a servant." Bruch in turn relayed this complaint to the War Department, adding that this lack of status and respect made it difficult to "keep good operators in the military telegraph service" and asked for relief from "the many abuses which operators are at present forced to submit." Mess arrangements were one of the leading complaints. Assistant Superintendent Captain William Fuller complained in his official report of September 1863 that officers often barred USMT operators from messing with them and relegated them to the enlisted men's mess. Captain Robert Clowry also had difficulty getting healthy drinking water for USMT operators at Hicks Station, Arkansas, because only officers were allowed to use the well at that camp. Other operators, like Luther Rose and William Hall, telegraphers with the Army of the Potomac, had decent quarters and mess arrangements. Rose and Hall usually had billets in private houses in camp or with officers on the march, and they often dined with their commanding general. On long marches, however, they too experienced the curse of all infantrymen: lice.[39]

An important reason that USMT operators felt entitled to status and respect was that they were skilled professionals who were entrusted with vital military information. Another reason was that many shared the same hardships and dangers as frontline troops. Of the USMT's 1,200 members, 25 died during their service, 8 as a result of enemy action. Another 175 were wounded or captured. Assistant Secretary of War Charles A. Dana recalled that at the Battle of Missionary Ridge, in November 1863, the USMT operators "kept at their post until the Confederates swept them out of the house. When they had to run, they went instruments and tools in hand, and as soon as out of reach of the enemy set up shop on a stump." With the "battle raging around them," they continued to operate. As Dana's account suggested, telegraphers were often exposed to enemy fire. For instance, USMT operator D. B. Lathrop died gruesomely in the Yorktown, Virginia, telegraph office in May 1862 because retreating Confederates had booby-trapped it with an exploding mine. Luther Rose, a frontline operator with the Army of the Potomac, came through the war unscathed, but he endured artillery fire on several occasions, including two near misses at Cold Harbor in June 1864 that killed a nearby mule and took off the camp provost marshal's leg. Rose mainly operated from the headquarters of the general commanding his corps and was further exposed to enemy fire because, when the headquarters moved, telegraphers were the last to move out.[40]

Linemen sent to repair downed telegraph wires faced similar dangers. In February 1864, for example, Captain William Gross sent two linemen out to repair wires downed by guerrillas in his Tennessee district in February 1864. They reported about twenty miles of wire down, with fifteen miles frozen into mud and unrecoverable. Unable to get a military escort, they set out anyway with four days' rations and forage for their horses. They returned unharmed. Two months later, however, four repairers in Clowry's Arkansas division were killed by bushwhackers, despite an escort of fifteen cavalrymen.[41]

Despite the hardships and dangers they endured and their importance to the war effort, members of the USMT were not eligible for postwar soldiers' pensions or membership in the Grand Army of the Republic (GAR). In response, former military telegraphers formed their own Society of the USMT, which lobbied for pensions and GAR recognition. Agitation for Federal veterans' benefits and for admission into the GAR continued into the twentieth century. Although the society regularly pressured Congress, collected testimonials from some twenty Federal generals, and continually petitioned the GAR for admittance, it did not succeed in achieving either goal.[42]

Besides these advocacy efforts, nostalgia for wartime adventure was another reason for the annual reunions of the Society of the USMT. As was the case with the GAR, reminiscences of shared wartime experiences were central to the wave of postwar nostalgia among military telegraphers. Like GAR veterans, military telegraphers at first expressed little interest in reunions. They did not hold their first reunion until 1882, the same year that William R. Plum published his two-volume celebratory history of the USMT. Plum himself was active in the formation of the Society of the USMT and actively solicited letters, diaries, and anecdotes from telegraphers to write his history.

Nostalgia was a regular feature of these reunions. At the 1885 reunion, for instance, attendees "laughed and chatted over the old times. . . . In fact, we swapped lies all day long, and smacked our lips for more." The 1886 reunion featured a chairman's gavel made by a former USMT telegrapher from wood from the battlefields of Chickamauga, Missionary Ridge, and Kennesaw Mountain, containing two embedded bullets. At the 1890 meeting, "relics of the war," including a box of bullets and shells labeled "Unsent Messages," graced the president's table. Telegraphers attending these reunions had become middle-aged and waxed nostalgic about their youth as carefree yet skilled practitioners of an important and arcane craft. Like Plum's 1882 history of the USMT, the reunion proceedings were replete with accounts of pranks and hard drinking.[43]

By the time the Society of the USMT held its first reunion in 1882, the commercial telegraph industry had become a near monopoly under the control of Western Union. One consequence of this industrial consolidation was a slow erosion of telegraphers' status. Postwar telegraphers earned less pay and had less workplace control than their wartime and prewar counterparts. Between the end of the war and the 1882 reunion, Western Union cut operators' wages from 25 to 40 percent, while expecting them to handle two to three times as many telegrams per shift. Operators responded by organizing trade unions and striking in 1870 and 1883, both of which strikes they lost.[44]

Operators thus hearkened back to the Civil War era as a golden age of telegraphing, when operators were respected and in demand. As operators endured wage cuts and work speedups during the 1870s and 1880s, many left the industry for better-paying occupations and consoled themselves with reminiscences and light fiction about this heyday. Tellingly, the Society of the USMT held its 1885 and 1886 meetings in conjunction with the Old Timers' Society, an organization of pioneer telegraphers. In April 1898 the society's members offered their services to President McKinley just before the start of war with Spain, a poignant yet empty gesture. At the September 1898 meeting, a former USMT superintendent complained that Army Signal Corps telegraphers had been "degraded to menials" and received only enlisted men's pay of eighteen dollars a month. The bitter strikes of 1870 and 1883 show that similar processes had eroded the skills and pay of commercial telegraphers as well as Signal Corps operators.[45]

The War's Effect on the Telegraph Industry

After twelve years of haphazard expansion and destructive competition, the telegraph industry entered a stable period in 1857 when the six leading telegraph companies (the American; New York, Albany, and Buffalo; Atlantic and Ohio; Western Union; South Western; and Illinois and Mississippi), followed in 1858 by the Montreal Telegraph Company, formed a cartel. The so-called Treaty of the Six Nations divided up the country into territories under the exclusive control of each company and set up an organization, the North American Telegraph Association (NATA), to discuss matters of joint concern and to resolve outstanding disputes. For instance, NATA set up common accounting and operating procedures that streamlined the transmission of messages over the lines of two or more companies, thereby eliminating vexatious errors and delays that characterized interregional telegraphing beforehand. The contract was also a defensive alliance to bar competitors from entering the industry after Morse's 1840 patent expired in 1861.[46]

While the telegraph companies all jockeyed for competitive advantage during the war, the six-party contract guaranteed their cooperation with each other and Federal authorities for military purposes. Indeed, their wholehearted cooperation stands in stark contrast to the disharmony between the commercial companies and the military in the Confederacy. At the beginning of the war, the War Department assumed nominal control over the commercial telegraph network but lightly exercised it only to censor press dispatches at Washington. Federal army officers never saw the need to commandeer commercial lines as southern generals had done in several instances.[47]

The commercial companies expected to be rewarded for their cooperation. As an 1863 NATA committee report phrased it, the companies "confidently" relied on "the enlightened sense of justice of the Government" and happily noted that "our confidence has been well repaid." The quid pro quo desired by the companies consisted of two major benefits. In return for giving priority to government and military traffic, the companies expected the government to help them earn great profits on commercial traffic. To this end, the USMT defrayed a large percentage of the expenses at joint offices. For example, the USMT paid 40 percent of salaries, rent, and cost of supplies at the Cairo, Illinois, office it shared with the Illinois and Mississippi and South Western Telegraph Companies. Also, the USMT allowed commercial companies to use military lines and operators to handle private messages when those lines and operators were idle, at no cost to the companies. Doing so, according to the NATA, had the added benefit of keeping USMT operators busy "when they would be tempted to leave their stations temporarily or spending time in idle conversation, over the wires, or discussing military movements about which they had better be silent."[48]

The extent to which these perquisites granted by the USMT boosted the profitability of the commercial companies is unclear. It is quite plain, however, that the companies profited enormously during the war. Even the South Western, which had its lines almost totally disrupted by the war, paid dividends of 12 percent in 1864. Its president, Norvin Green, noted in October 1864 that, despite the "utter destruction of the central portions of our lines, and the uncertain condition and status of the southern portions," the company's profits were "larger and more uniform and reliable than they have ever been." Similarly, the American, which lost its southern lines at the outbreak of the war, about a third of its network, paid its shareholders quarterly dividends of 2 percent from 1861 through 1863, raised to 2.5 and 3 percent quarterly in 1864 and 1865. The Illinois and Mississippi fared even better. Its receipts tripled during the war, and it consistently earned above 30 percent over expenses.[49]

Western Union fared the best. In 1863 and 1864 alone, the company paid cash dividends of 42 percent and stock dividends of 200 percent. Residents of Rochester, the company's headquarters, reportedly sold their furniture and mortgaged their houses to buy its stock. By 1864 the company's capital stood at more than $21 million, representing a grossly inflated capitalization of about $485 per mile for a network of about forty thousand miles of line. By comparison, Western Union's largest rival, the American Telegraph Company, operated some twenty-three thousand miles of wire at a far lower capitalization of $2 million. Standard construction costs at the time were $100 to $150 per mile for a single-wire line. Western Union executive James D. Reid later recalled that the two wartime stock dividends totaling 200 percent were "clear and unmixed water" that "seriously endangered the stability of the company." Managers of other telegraph lines were well aware that Western Union's capitalization was grossly inflated, a sticking point for them during the merger negotiations of 1866.[50]

The second part of the quid pro quo expected by the telegraph companies was that after the war the government would turn over military lines built within their territories at nominal cost. The NATA justified this in 1863 by claiming that many of the military lines were poorly located for service as paying commercial lines and thus probably only valuable as salvage. When the USMT disbanded in February 1866, it turned over all of its lines south of the Ohio River, some fourteen thousand miles, to the American and South Western companies merely in consideration of any claims those companies might have against the government for the wartime use of their lines and offices. The commercial companies paid nominal sums for military lines built north of the Ohio River. For example, the Illinois and Mississippi Telegraph Company paid the federal government $15 to $25 per mile for several hundred miles of lines in Illinois and Missouri. During the war the company had charged the government $65 to $100 per mile for building the very same lines.[51]

Such hand-in-glove cooperation between the USMT and the commercial lines was no accident. After all, the managers of the USMT were usually also employees of the commercial companies. Gross and Van Duzer, for instance, were simultaneously assistant superintendents of the USMT and the South Western Telegraph Company. Their immediate superior in the USMT, Colonel Anson Stager, remained in Cleveland, Ohio, as superintendent of the Western Union Telegraph Company. These dual roles often led to conflicts of interest, particularly after the end of the war when the USMT turned over all its property to the commercial lines at bargain-basement prices.

Stager's appointment was especially controversial among the managers of other telegraph companies. Despite his role as superintendent of the whole USMT, he continued to concentrate on his civilian position as superintendent of Western Union at Cleveland. He left much of the day-to-day responsibility of running the USMT in the hands of his assistant superintendents and Thomas T. Eckert, a Western Union manager stationed in Washington. Industry insiders suspected that Stager placed Western Union's and his own financial interests ahead of the country's. Thomas Bassnett, treasurer of the Illinois and Mississippi Telegraph Company, claimed that Western Union had used its "political influence" to place Stager at the head of the USMT to improve its competitive position within the industry. According to George Kennan, a young Western Union operator, it was common knowledge among telegraphers that Stager had made a fortune speculating in gold, since he was "the first man in the U.S. to know the news" and placed trades in New York before other speculators learned the news. George Ladd, a telegraph superintendent in California, similarly made his

fortune speculating in gold. Kennan also claimed that Stager deliberately recruited the top operators for Western Union's offices and shunted inferior or
troublesome operators into the USMT. In fact, Kennan asked Stager twice to
transfer him from Western Union to the USMT, only to be told that he was too
valuable to leave.[52]

Other USMT officials exhibited similar patterns of behavior. William Gross
and John Van Duzer, for example, served as superintendents of the USMT in
1865. Their jurisdictions covered much of the territory of the South Western
Telegraph Company, for which they also served as assistant superintendents. In
the summer of 1865, as Norvin Green worked to reestablish that company's control over its lines in the former Confederacy, he complained to both men about
the heavy expenses the company had incurred in service to the Federal military.
Green told Gross that he had reached an understanding with Stager and Eckert
that division superintendents had "entire authority and discretion" for reimburs-
ing the South Western's expenses. Green discouraged Gross from giving government officials "reasons or explanations" about reimbursement, "which only serve
to elicit enquiries and cause postponements, embarrassing the settlement of accounts." On one occasion, Green wanted Gross to confiscate the lines of the Arkansas State Telegraph Company, which had cooperated with the Confederacy,
and to turn the lines over to the South Western without officially notifying his
military superiors: "There being no official record of the seizure, there is in fact
no seizure; and you need not make any, and there is an end to it." For their cooperation, Green amply rewarded both men. Green gave Gross $2,000 of South
Western stock at par, when it traded at $120, telling Gross that he could pay for
it "in such sums as from time to time you may wish." Green also brought Gross
to Western Union's New York headquarters as a junior executive after the great
mergers of 1866. It is unknown if Van Duzer also received a gift of stock from
Green, but Green promised him management of the line South Western planned
to build to the Pacific.[53]

The actions of these telegraph managers complicate our understanding of
Civil War procurement. The Federal government spent about $1.8 billion in 1860
dollars during the war, more than all previous expenditures combined. Two-
thirds of this spending directly outfitted and provisioned the army and navy.
Such a vast amount of spending by an ill-prepared bureaucracy led many contemporaries (and later historians) to assume that military procurement was either a matter of party patronage or unfettered capitalism; yet most of the military
personnel and the civilians involved in the procurement effort discharged their
duties faithfully.[54] Indeed, USMT personnel on the whole patriotically served the

Union war effort with little conflict of interest between their duties to country and company. Yet telegraph managers often exploited their USMT positions for personal gain or to advance the interests of their companies.

The Civil War disrupted the stability of the six-party contract by severing the lines of the American along the Eastern Seaboard and throwing the affairs of the South Western into chaos. Of the six major telegraph companies, Western Union benefited most from the dislocations of the war. Indeed, the great telegraph consolidations of 1866, culminating in a national telegraph network under Western Union's control, was a direct consequence of the war. But the outcome might very well have been different if the Atlantic cable of 1858 had continued working, or if the 1865 Atlantic cable expedition had succeeded. Had there been a working Atlantic cable by the end of 1865, the American Telegraph Company would have likely dominated the industry instead. After all, many of the investors in the cable project were also directors and major stockholders in the American, and all cable traffic would have had to pass over that company's lines.

Nevertheless, a working Atlantic cable entered service in the summer of 1866, just months after Western Union had bought out the American, its last remaining rival. Under the aggressive leadership of Hiram Sibley, Western Union exploited the war years to achieve a near monopoly in telegraphing. Thomas Bassnett, the treasurer of the Illinois and Mississippi, suspected as early as December 1861 that Western Union had "had a grand and far-reaching policy calculated in time to swallow up or make tributary all other organizations." Sibley's actions soon confirmed Bassnett's suspicions. Sibley's first two takeover targets were the smallest and most vulnerable of the cartel of six companies, the New York, Albany, and Buffalo and the Atlantic and Ohio companies. The former company connected New York City to the Great Lakes and the latter ran across Pennsylvania between Philadelphia and Pittsburgh. Both lines were important acquisitions for Western Union, because its lines did not yet reach the Eastern Seaboard. After acquiring them in early 1864, Western Union gained entry into both New York City and Philadelphia, placing Western Union squarely in the heart of the American's territory.[55]

Another key reason for Western Union's success was the completion of its line to the Pacific coast, which began operating in October 1861. Hiram Sibley grandly envisioned the California line as the first link in a vast project linking the European telegraph network with the United States, through Asiatic Russia, Russian North America, and British Columbia. In the wake of several failures (in 1857, 1858, and 1865) to lay and operate an Atlantic cable, the so-called Russian Extension seemed to be the most promising way to connect the hemispheres. The project was the brainchild of Perry McDonough Collins, an American eager to secure the commerce of Siberia and China for the United States. Sibley embraced Collins's project wholeheartedly, reasoning that if Western Union could beat the Atlantic cable project in linking the continents, his company might well become the dominant telegraph company on the globe. However, Western Union abandoned the Russian route in early 1867 after the Atlantic cable had operated successfully for several months and after Western Union had acquired the American Telegraph Company.[56]

While the Russian Extension remained unfinished, Western Union's line to California had more immediate and concrete purposes. It was, of course, highly profitable. Western Union built the line for $500,000, and it took in $460,000 in federal subsidies alone in the decade after its completion. Its rates also ensured that the line would pay its investors well. To telegraph ten words to California, customers paid $5 from Chicago and $6 from New York and Washington. For Sibley, Wade, and other Western Union insiders, the line gave them opportunities for stock jobbing on an unprecedented scale. To finance the line's construction, estimated at $147,000 for the section from Nebraska to Salt Lake City, the Pacific Telegraph projectors issued $1 million in stock, later exchanged for $2 million of Western Union's stock. In 1863 and 1864, Western Union issued two stock dividends of 100 percent each, thus tripling the company's capitalization. So for an expenditure of $147,000, the telegraph line to the Pacific earned its promoters $6 million of Western Union's stock. Furthermore, the completion of the line gave Western Union a great deal of prestige, particularly when coupled with the company's global ambitions to build a line through Russian America and Siberia.[57]

The final factor in Western Union's rise to dominance was the expiration of Morse's 1840 telegraph patent in 1861. This threw the industry open to competition. Although telegraph designs other than Morse's had entered service in the 1840s and 1850s, in particular Alexander Bain's chemical telegraph and Royal E. House's printing telegraph, these had quickly proved to be more expensive or less reliable than the straightforward Morse key and sounder.[58] Thus, the 1861 expiration of Morse's patent removed a major barrier to entry into the telegraph market. Within a few years, three new companies began to compete with the 1857 cartel, the Bankers & Brokers, Insulated, and Independent. The Independent proved the most formidable. After it was renamed the United States Telegraph Company (USTC), its managers announced their intention to build a nationwide telegraph network, including a line to California to rival Western Union's and an Atlantic cable via Labrador and Greenland. By 1865 the company's future seemed

assured. Its revenues, wire miles, and offices were increasing weekly. It had se-
cured key contracts with major railroads to build on their rights-of-way and had
just paid a 4 percent dividend to its shareholders. However, industry insiders
were aware that the company's performance was too good to be true. Its stock
dropped from par in March to $20 in October. At this point, the line's president
resigned, and its directors hired William Orton, then commissioner of the Internal Revenue Bureau. When Orton took over the USTC in December 1865, he was
at first cautiously optimistic, telling one investor that the company stood a fair
chance to compete with Western Union, as long as it remained under "judicious
management." But Orton soon realized that the company's lines were shoddy and
that it was not earning its expenses. By January 1866 he favored a consolidation
with Western Union as the only way to staunch the flow of red ink.[59]

Western Union, for its part, was content to sit back and watch the USTC lose
money. Sibley, Wade, and Western Union's managers thought at first that the
company was merely a scheme for fat construction contracts and stock jobbing.
Even so, they reasoned, the upstart company hurt its rivals more than Western
Union. A prominent New York investor in the Illinois and Mississippi wrote its
president John Caton in March 1865 that his company "would be pretty much
sandwiched" between Western Union's Pacific line and the growing United
States' network, and that Caton's only "salvation" was consolidation with Western Union.[60]

Western Union consolidated with USTC in March 1866. In the meantime, the
American had bought out the South Western, leaving only two rivals to Western
Union, the American and the Illinois and Mississippi. By any measure, Western
Union's directors overpaid to acquire the United States. United States stockholders received a quarter of the stock of the consolidated company, even though the
ratio of their earnings was about one to seven. In addition to the desire to remove
a troublesome competitor, Western Union's directors were willing to overpay to
acquire the United States for another reason: they could use control of its lines
to force the American and Illinois and Mississippi into a grand consolidation.
By the terms of the six-party contract, Western Union was forbidden to operate
lines in the territories of its rivals. USTC was a rival to the cartel and operated its
lines wherever it pleased. Thus, by acquiring the United States, Western Union
was in fact operating lines in its rivals' territories. Western Union expressed its
willingness to sell off the United States lines in the American's and Illinois and
Mississippi's territories, but its managers could never quite agree on a price with
the other two companies. As negotiations dragged on, Western Union continued
to operate the lines to the detriment of its rivals.[61]

Western Union's acquisition of the United States Telegraph Company left its two remaining rivals with two options. They could attempt to build a telegraph network of national scope to compete with Western Union, likely resulting in protracted, costly, and uncertain competition, or they could consolidate on advantageous terms. They opted for the latter course. Although the Illinois and Mississippi maintained its independence as a regional telegraph company until June 1867, the directors of the American Telegraph Company voted to merge with Western Union in June 1866. After this merger, Western Union moved its headquarters from Rochester to the American's New York headquarters on Broadway.

The past five years had dramatically changed the American telegraph industry. No longer a cartel of six regional companies, by the summer of 1866 it had become a near monopoly under Western Union's control. Western Union itself had evolved from a regional telegraph line headquartered in upstate New York to an integrated system spanning the continent, inheriting in the process a Broadway office building befitting such a company. Over the next several years, ownership and control moved from a small group of Rochester and Cleveland insiders headed by Hiram Sibley and Jeptha Wade to wealthy New York capitalists like Cornelius Vanderbilt.[62] By 1870, the telegraph entrepreneurs of the pioneering generation had given way to the so-called robber barons who would come to symbolize the rapacity of late nineteenth-century American capitalism.

CHAPTER TWO

"As a Telegraph for the People It Is a Signal Failure"

THE POSTAL TELEGRAPH MOVEMENT

When Samuel Morse brought his invention to the attention of the secretary of the treasury in 1837, he wrote: "The *mail system* . . . is founded on the universally admitted principle, that the greater the speed with which intelligence can be transmitted from point to point, the greater is the benefit derived to the whole community." It was therefore "most natural" to him that the Post Office should operate his telegraph since it was merely "another mode of accomplishing the principal object for which the mail is established, to wit: the rapid and regular transmission of intelligence." Several years later, Morse renewed his call for postal control for another important reason, that his telegraph might become "an engine for good or evil" if left "in the hands of private individuals or associations."[1]

When Morse obtained congressional funding in 1843 to build a demonstration line between Washington and Baltimore, few questioned that the federal government ought to develop this new technology. After all, only the state had sufficient capital and the necessary will to construct a national network—the undertaking appeared too immense for private capitalists. Telegraphy also fit well with a long tradition of government sponsorship of internal improvements.[2] From a constitutional perspective, it seemed that this new form of communication properly belonged within the Post Office Department's authority to "establish post roads and post offices." Indeed, at the time all national governments owned and operated the major channels of communication within their borders, including the existing optical telegraph networks of France, Russia, and Sweden. Only England allowed a privately owned electrical telegraph system to operate between 1839 and its nationalization in 1870. As later postal telegraph advocates continually pointed out, the United States remained an anomaly among the world's nations by permitting private ownership of electrical communication.

The Post Office Department owned the pioneer Washington-to-Baltimore line from May 1844 until the Magnetic Telegraph Company bought it in March 1847.

During those three years Postmaster General Cave Johnson (who had ridiculed Morse's project while serving in Congress) repeatedly urged Congress to place the telegraph under permanent postal administration. Echoing Morse, Johnson warned in 1846 that "an instrument so powerful for good or for evil, could not . . . be permitted to remain" under private ownership, and that the Constitution granted the Post Office Department "exclusive power for the transmission of intelligence." Besides, Johnson feared, the telegraph might come to supersede the mails entirely. For these reasons, he insisted that the government could brook no competition from a privately owned telegraph.

Despite Morse's and Johnson's appeals, the federal government abandoned the telegraph to capitalists. They had trouble persuading Congress to retain Morse's line to Baltimore because it did not earn enough to pay its expenses. This was hardly surprising, since the line was an experiment and Baltimore was only about forty miles away. Additionally, the federal government in this period was remarkably reluctant to take on new responsibilities.[3] Although the line to Baltimore was not profitable, by 1846 private investors had enough faith in Morse's invention to finance the construction of several commercial lines between Boston and Washington, with new lines planned westward to the Mississippi and Great Lakes and southward to New Orleans. Morse's legal and financial advisers, Amos Kendall and Francis O. J. Smith, had come to realize that a private-sector telegraph was far more lucrative in the long run than selling out to the government.[4]

By 1860, then, most Americans had come to accept the permanent presence of a privately owned telegraph network that served two principal classes of customers, the press and businessmen. Few citizens saw the need for government control. Instead, the question of federal involvement in the industry revolved around the desirability of subsidizing particular projects such as lines to the Pacific coast and a transatlantic cable. But two consequences of the Civil War converged to reanimate sentiment for government ownership. The war accelerated an existing trend toward consolidation in the telegraph industry. In 1866 Western Union bought out its last two major rivals. Afterward, it would control at least four-fifths of the nation's telegraph traffic. Western Union's 1866 consolidation and its continued dominance of the industry alarmed Americans across the political spectrum. This consolidation coincided with the rise of an activist federal government during the Civil War and Reconstruction.[5] Many Americans wished to see this newly energized state break up, or at least weaken, Western Union's control over the nation's electrical communication network.

The postal telegraph movement unfolded in several stages between the end of the Civil War and World War I.[6] The movement began in 1866 with the passage of the National Telegraph Act, the first major law to regulate the communications industry. The most important provision of this bill laid out a mechanism for Congress to buy out private telegraph companies any time after 1871. In 1868 attention shifted to Boston businessman Gardiner Hubbard's proposal to secure a congressional charter for a new company that would contract with the Post Office to transmit low-cost telegrams. Hubbard's plan occupied the attention of both Congress and Western Union for about the next five years. The postal telegraph movement was at low ebb between the mid-1870s and 1881 because of renewed competition in the telegraph industry, thanks to two companies backed by financier Jay Gould. In 1881 it became clear that Gould had used these companies as stalking horses to gain control of Western Union, ending telegraph competition for several years. Thus, from 1881 to the early 1890s, reformers renewed their calls for postalization, mainly because of Gould's control of Western Union and a bitter telegraphers' strike in the summer of 1883. The postal telegraph movement gained broader support during the 1890s as both the Populist Party and American Federation of Labor adopted it as key planks in their legislative agendas. From about the turn of the century until 1913, however, the issue disappeared from public debate. Even a 1907 telegraphers' strike did little to renew interest. The final stage of the postal telegraph movement occurred during the Wilson administration. Many members of Wilson's cabinet were strong advocates of public ownership of utilities. Almost immediately after taking office they began considering a government takeover of the telegraphs and telephones. In 1918 Postmaster General Albert Burleson obtained control of the wires, claiming wartime necessity as the reason. Yet Burleson and other administration officials admitted that they wanted government control to be permanent. By 1919 it had become clear that government operation of the wires was a disaster, and under pressure from the public and a Republican Congress, the administration restored private control. This episode ended the postal telegraph movement for good and helped to kill off support for public ownership of utilities generally.

The fifty-year debate over Western Union's dominance of the telegraph industry helped shape modern politico-economic thought, particularly with regard to the role of the state in regulating networked technologies. More specifically, the postal telegraph debate posed and clarified three important issues. An industrial monopoly of national scope was a new feature of the American economic landscape. Reformers sought to understand why this had occurred and, more importantly, how Western Union was able to retain its supremacy well after 1866. At first glance, the telegraph was one of the first examples of what economists would come to call a natural monopoly, an industry that tends toward consolidation because of intrinsic properties of the technology deployed, such as its capital costs, marginal cost structure, and presence of network externalities. However, the perpetuation of Western Union's dominance after the Civil War had as much to do with the company's strategy as with ineluctable laws of economics. The company's continued ascendancy was by no means inevitable, and it faced determined competition at key points in its history. Western Union used hardheaded business practices to fend off competitors, particularly exclusive rights-of-way contracts with several hundred railroads and exclusive contracts with the nation's major press associations.

Antimonpolists regarded an untrammeled Western Union as a serious threat to the nation's political and economic welfare. In particular, they feared the consequences of Western Union's control over the flow of the nation's political and financial news, especially after Jay Gould acquired control of the corporation in 1881. Of more lasting significance, the postal telegraph debate hinged on a deeper question, whether the telegraph ought to remain an expensive business medium or become a social medium like the postal system and the British Post Office telegraph, affordable enough for ordinary citizens to use. Critics of Western Union routinely complained that the company charged extortionate rates in order to pay lavish dividends on its inflated capital. They contended that the government could build and operate a national telegraph network for a fraction of Western Union's capitalization. According to this logic, such a system could charge far less than what Western Union charged and would make telegraphing a popular medium. Western Union's defenders countered that the telegraph was properly an expensive business medium whose major customers prized speed over cost. Furthermore, they contended, the true way to value Western Union's worth was not its plant cost but its earning potential. Thus, the debate over telegraph rates helped hammer out a second concept important to both corporate finance and public utility economics, how to measure the true value of a corporation. Traditionally, corporations were expected to maintain a capitalization close to the value of a company's physical plant. Shareholders expected to profit from their investments mainly through dividends instead of rising share prices. Thus, corporations whose capitalizations greatly exceeded their book values were accused of stock watering, a term that implied fraud and manipulation. However, investors and managers in the last third of the nineteenth century would come to rely on a company's future earnings potential as a more accurate way to assess a company's market capitalization and share price. On the other hand, public-utility regulators after the turn of the century would adopt a yardstick similar to that of telegraph reformers, setting rates that were based on a fair return on the value of a utility's physical plant.

Reformers and legislators could never agree on a specific policy to contend with the telegraph monopoly, although they considered a plethora of options. The general reason for their failure to coalesce around one of those options was their uncertainty about how the state ought to oversee natural monopolies like the telegraph industry. As it evolved, the postal telegraph movement became an important seedbed for a larger public ownership movement of the late nineteenth and early twentieth centuries, particularly the municipal ownership movement. Economists and reformers such as Henry George, Edward Bellamy, Richard T. Ely, Edwin R. A. Seligman, Simon Newcomb, and Frank Parsons all placed postal telegraphy within a larger public ownership program. While the municipal ownership movement met with a degree of success, the postal telegraph movement was largely a failure. Although the movement exerted quasi-regulatory pressure on Western Union to reduce rates and improve service, reformers were unable to achieve their main objective, a nationalized telegraph network, until the First World War. Government operation of the wires in 1918 and 1919 proved disastrous and was instrumental in dampening enthusiasm for public ownership at the state and federal levels. Thereafter, public ownership remained limited to utilities on the municipal scale, and regulation became the preferred method for managing utilities of larger geographic scope.

Western Union and the Problem of Natural Monopoly

Western Union became the nation's first industrial monopoly in 1866 as the culmination of an existing trend toward consolidation in the telegraph industry. The industry's monopolistic structure, whether as a contract-based cartel between 1857 and 1866 or dominated by a single firm afterward, was due to telegraphic communications being a so-called natural monopoly, one of the three major forms of monopoly in the American economy in the nineteenth century. The oldest type relied on special privileges or protections granted by state or federal authorities. Jacksonians of a generation before had railed against these sorts of monopolies, the Livingston steamboat monopoly and the Second Bank of the United States being leading examples. After the Civil War, few monopolies of this archaic type remained at the state or national levels, thanks mainly to state general incorporation laws. The large industrial combine was the quintessential monopoly of the late nineteenth and early twentieth centuries. These much-vilified "Trusts" established and maintained themselves through predatory pricing, horizontal and vertical integration, shipping rebates, and interlocking directorates. For contemporaries, Standard Oil was the most notorious example of this sort of monopoly.

The telegraph industry was one of the earliest examples of a new form of monopoly, what economists would come to call a natural monopoly. The term describes a situation in which only one firm can produce profitably in a particular industry because of high barriers to entry, increasing returns to scale, or the presence of network externalities. Although the term originated with British economist and philosopher John Stuart Mill in the 1840s, American economists, particularly Richard T. Ely and Henry C. Adams, gave it full articulation in the late 1880s. In theory, natural monopolies typically occur in industries using networked technologies and offering a standardized product or service. In practice, however, natural monopolies are far from inevitable or permanent. Rivals can successfully compete with an established incumbent by offering a differentiated product or service or by adopting methods resulting in a different fixed or marginal cost structure. Thus, the ability of an incumbent to retain its dominance depends to a large extent on the technological state of the art in that industry.[7]

In the case of the telegraph industry after 1866, construction and equipment costs were relatively low and the fundamental Morse patents had expired. These low barriers to entry emboldened several entrants to compete with Western Union. Nearly all failed within a few years because they were unable to sustain competition on a national scale. This was partly due to what economists today call network externalities. This principle, an intrinsic feature of point-to-point communication networks, holds that the value of a network to each individual user rises as more users are connected. For nineteenth-century customers, the telegraph became more useful as it expanded to become a national network reaching virtually the entire country's population. Thus, Western Union's combination of geographic reach and network density was perhaps the highest barrier to would-be competitors. Upstarts could succeed over the long run only if they constructed a national network that competed with Western Union for the bulk of telegraph traffic. In practice, some lines managed to compete for a time on the Eastern Seaboard or between particular commercial cities, but they could not offer their customers access to thousands of offices and tens of thousands of miles of line. As Richard T. Ely succinctly put it in 1887, "One telegraph company can send telegrams all over the country cheaper than two; hence the absorption of all companies . . . by the Western Union."[8]

The attempt of the United States Telegraph Company to compete with the established telegraph cartel between 1864 and 1866 demonstrated this concept clearly. The United States company attempted to build a network that spanned the entire country, including a line to the Pacific coast. In November 1865 the company hired William Orton, head of the U.S. Internal Revenue Commission, as its president. Orton quickly realized that competition was ruinous; the United States lines were losing $10,000 a month. He immediately entered into merger negotiations with Western Union, and at the end of February 1866 he obtained the best terms he could get for his shareholders. Just after the merger with Western Union, Orton explained to one investor that attempting to compete was futile: "The Co. never made a dollar—and the further its lines extended the more it lost." Similarly, in 1884 Thomas Edison and telegraph executive Erastus Wiman dismissed the possibility of an upstart company profitably competing with Western Union for the cream of the telegraph business, traffic between the nation's sixteen largest commercial centers. They concluded that this traffic accounted for only 13 percent of Western Union's receipts and that such a company would be "a most dismal failure."[9]

Although economists would not fully articulate the meaning of "natural monopoly" until the late 1880s, the concept was quite familiar to both telegraph industry executives and critics as early as 1866. Early that year, Senator B. Gratz Brown introduced a bill to establish a Post Office telegraph because "there is no competition now in the telegraph system, and cannot be any unless the Government interposes." In reply, the presidents of the American, United States, and Western Union companies (close to completing negotiations for consolidation) objected to Brown's proposal to build a competing government telegraph network. They agreed with Brown that competition was impossible in telegraphy, particularly when attempting to "cover the country." However, they claimed that their consolidation was not truly a monopoly because it was "the result, not of power, but of the pressure of mutual necessity" to avoid ruinous competition. Similarly, in its annual report for 1878, Western Union told investors that its continual modest rate reductions "demonstrate that one telegraph company, . . . relieved from the expense of numerous competing offices, can afford to do the business of the country at lower rates, with a living profit," than could two or more competing companies. In 1881, just after Jay Gould obtained control of Western Union, an anonymous pamphleteer told prospective investors that Western Union was a safe stock because competition in telegraphy "has never been permanently successful. . . . *In the nature of things this is impossible.*" Similarly, Bell Telephone Company president William H. Forbes refused in 1883 to deal with competing telegraph companies because of their "temporary character." Forbes regarded these rivals as stock-jobbing schemes set up for the sole purpose of getting bought out by Western Union. "The footsteps of all lead into the Western Union cave; not one has ever returned," Forbes concluded.[10]

Yet Western Union's continued dominance was not inevitable, and the company expended considerable energy to maintain it. Because of low construction costs and the expiration of Morse's fundamental patents, network externalities were not enough to bar competitors from trying their luck. Western Union relied on four tactics to best rivals: exclusive right-of-way contracts with several hundred railroads, exclusive contracts with press associations, punishing rate wars, and control of key technologies. The Atlantic and Pacific Telegraph Company told its shareholders in 1877 that it had agreed to a rate pool with Western Union after several years of fierce competition because it could no longer withstand "the contest . . . over rights of way, contracts of connection; press, market and other special service; office locations in hotels, commercial exchanges, and other public resorts; and for patent rights and many other interests."[11] Only one competitor, the Postal Telegraph Company, was able to stay in business on a sustained basis, from the mid-1880s until World War II. Postal did so only by accepting a junior position in the industry. Postal also had a different business model from Western Union's—it viewed its landlines as feeders for its overseas cables and chose to compete with Western Union only at commercial points that generated significant cable traffic.

In 1912 Western Union attorney Henry D. Estabrook dubbed the railroad and telegraph the "Siamese twins of Commerce" because of their close and historic relationship. Estabrook also had in mind Western Union's most important competitive advantage, its exclusive rights-of-way contracts with nearly all the nation's railroads. It was no accident that the railroad tycoons Cornelius and William H. Vanderbilt, Jay Gould, and Robert Garrett were all active in the telegraph industry. Indeed, the battle between the Vanderbilts and Gould for control of Western Union in the 1870s and 1880s and the Baltimore and Ohio's intense but short-lived competition with Western Union in the mid-1880s were but additional fronts in the railroad wars between these men. In the mid-1850s telegraph companies began to sign contracts with railroads giving the telegraph line an exclusive right-of-way along railroad lines in exchange for giving railroads free telegraph service for train dispatching and other purposes. For telegraph companies, this rendered construction and maintenance far less expensive than building on public highways. By 1881 Western Union had such contracts in place with some eight hundred railroads. Jay Gould told the Senate Committee on Education and Labor in 1883 that these contracts made it "perfectly impracticable" for another telegraph company to compete effectively with Western Union. For example, the

Pacific and Atlantic Telegraph Company was unable to find a railroad right-of-way into New Orleans in 1871 because of Western Union's exclusive contracts, despite offering $350 per mile (half in cash, half in stock) for similar rights-of-way elsewhere. Similarly, the Atlantic and Pacific Telegraph Company told its shareholders in 1875 that Western Union's attempts to prevent it from gaining access to a key railroad right-of-way would "delay the completion of that line, increase the cost of construction, [and] occasion a loss of business." The only option for Western Union's competitors was to sue in state courts to condemn railroad rights-of-way under eminent-domain law, a lengthy and expensive process with no guarantee of ultimate success. The Postal Telegraph Company spent two years in state courts in the 1890s to condemn rights-of-way along the Southern Railway and Mobile and Ohio Railroad.[12]

Competitors frequently protested that the National Telegraph Act of 1866 prohibited contracts for exclusive rights-of-way. That certainly seemed to be the intent of the framers of the act, to foster competition by making it easier for rival telegraph companies to gain access to these rights-of-way. For a few years after the act's passage, some state courts interpreted it this way. Yet federal courts ultimately did not. Instead, they held that it merely prohibited state legislation from interfering with rights-of-way contracts and that telegraph companies still needed the permission of railroads to build lines alongside their tracks. In other words, federal courts routinely upheld exclusive rights-of-way contracts between railroads and telegraph companies, leaving rival telegraph lines the expensive option of condemnation proceedings in state courts. The U.S. Supreme Court issued a definitive and comprehensive ruling on exclusive rights-of-way contracts in 1904, when it upheld them in *Western Union Telegraph Company v. Pennsylvania Railroad Company*. Armed with this ruling, Western Union continued to rely on exclusive rights-of-way contracts for the vast majority of its pole mileage. In contrast, Postal was able to obtain only a handful of such contracts.[13]

Because of Western Union's dominance in press business and ticker service, competitors were unable to generate significant business from newspapers and brokers, collectively the heaviest users of the telegraph. To borrow the language of electric utility economics, these served as Western Union's "base load," a class of business that provided constant and reliable income from year to year. Western Union's exclusive contract with the New York Associated Press (NYAP) was of prime importance in denying rivals an important source of business. In 1872, for instance, NYAP provided more than 200 daily newspapers with dispatches, all sent over Western Union lines. By 1880 this had increased to 355 dailies. By contrast, NYAP's rival the American Press Association (APA) served 84 dailies in 1872. The APA was an ally of Western Union's major competitor at the time, the Atlantic and Pacific Telegraph Company. Although Western Union did not enjoy exclusive contracts with boards of trade and stock exchanges, it did insist on exclusivity in providing ticker service and leased wires to brokers who traded on those exchanges. Brokers who leased wires from another telegraph company or from the Bell Telephone Company found themselves cut off from Western Union's ticker service, their instruments and wire connections physically removed.[14]

Western Union also beat down competitors through vicious rate wars. Competitors underwent a similar life cycle. After a rival appeared, it and Western Union would lower rates drastically at points where they competed until the rival was forced to the wall. Western Union then bought out or leased its lines or in a few cases agreed to a rate pool. Thereupon rates returned to the level they had been before the new competitor had appeared. Western Union was able to use its financial muscle and extensive geographic scale to absorb losses at competing offices until its competitor gave in. The Postal Telegraph Company was the one exception to this pattern. The company was able to withstand a rate war with Western Union in both the landline and cable businesses between 1885 and 1888 because Nevada silver titan John Mackay covered its losses until Western Union gave in. During the cable rate war, Western Union slashed its rates from seventy-five cents per word to twenty-five cents, which the Mackay-Bennett cables matched. When Western Union dropped its rate to twelve and a half cents, Mackay refused to follow. Instead he appealed to his customers' long-term interests, telling them that, if Western Union won the rate war, it would immediately raise its rates back to seventy-five cents. The appeal worked, and Western Union decided that it could no longer withstand yearly losses of three-quarters of a million dollars in its cable business. Jay Gould reportedly exclaimed that it was impossible to beat Mackay: "If he needs another million or two he goes to his silver mine and digs it up."[15]

Throughout the last two decades of the nineteenth century, Western Union's critics routinely charged that the company had deliberately retarded the pace of technological progress in order to bar competitors from entering the industry. Western Union's technological conservatism accelerated after William Orton's death in 1878 and especially after Gould's takeover in 1881. Thomas Edison, for example, later recalled that when Gould obtained control of Western Union, "I knew that no further progress in telegraphy was possible and I went into other lines." In 1883 Henry George testified to Congress that Western Union paid electricians "*not* to invent" and that the company acquired "invention after invention simply to hold them idle." By the turn of the century, this view had become axiomatic among electrical engineers. At the Franklin Institute's 1898 meeting, Reginald Fessenden remarked that it was "generally understood" among engineers that Western Union and Postal "do not want any improved apparatus" and that they bought patents only to prevent potential competitors from adopting them. In 1902 Michael Pupin charged that Western Union and Postal provided the public with lackluster and expensive service because they used "antiquated methods." Neither company, Pupin claimed, spent even "10 cents a year for experiments." The editors of the *Electrical World* agreed with Pupin, calling telegraphy a "stagnant industry" and urged the companies to improve telegraphy so that "instead of sending one message a year, the average American will send ten." In response, Western Union engineer John C. Barclay sidestepped the issue of poor service for regular telegrams and merely defended the excellent and rapid service his company gave stockbrokers.[16]

Fessenden's and Pupin's criticism had much merit. Although Western Union executives rarely stated this publicly, they knew that a static state of the art perpetuated the company's monopoly. After all, the company adopted only two important technological improvements between the Civil War and turn of the century, Edison's quadruplex and the stock ticker, both between the late 1860s and mid-1870s, and the company remained wedded to the traditional Morse key and sounder until after 1910. In contrast, some of Western Union's competitors eagerly embraced technological innovations. The Atlantic and Pacific, for instance, relied on a high-speed automatic telegraph designed by Thomas Edison to handle press dispatches for the American Press Association. Western Union manager James D. Reid credited this system with giving the company a decided advantage in handling press traffic.[17]

As might be expected, Western Union attempted to control key technologies in telegraphy. In several cases the company acquired patents merely to prevent competitors from using them. In 1872 Orton sought the American rights to an automatic rapid printing telegraph capable of transmitting several hundred words per minute invented by the celebrated British electrician Charles Wheatstone, despite his doubt that it "possesse[d] any value" to the company. Instead, he wanted to forestall rivals like the Atlantic and Pacific from using it. Western Union acquired the rights in 1874 but did not install a set for seven more years. Despite its widespread use by the British Post Office, Western Union had deployed only sixteen Wheatstone sets by 1886, as against some thirty-two thousand Morse instruments in use. Similarly, Western Union attempted to cover the field of multiplex telegraphy by hiring electricians like Thomas Edison to invent and patent as many variations as possible to prevent rivals from using duplex or quadruplex transmission methods.[18]

In 1853 British railway engineer Robert Stephenson famously told Parliament, "Where combination is possible, competition is impossible."[19] Although he was speaking of English railway consolidations, his remarks applied with equal force to the American telegraph industry. Profitable and sustained competition with Western Union proved impossible after 1866 (except for Postal Telegraph Company's unique focus on its overseas cables). While Western Union's anticompetitive strategies called into question how "natural" its monopoly was, it was clear that the industry tended toward consolidation because of intrinsic technological and economic characteristics of telegraphy itself. Thus, the telegraph industry posed a novel and unique problem. Other forms of monopoly, those based on special governmental privileges or those resulting from large accumulations of industrial capital, lent themselves to legislative or free-market solutions. However, no mechanism yet existed to regulate the rates and terms of service of a telecommunications company spanning the entire country. Interstate telegraphy would not fall under federal regulation until the passage of the Mann-Elkins Act in 1910. Furthermore, any serious and sustained competition to Western Union seemed destined to fail. Thus, during the last third of the nineteenth century, the only two long-term alternatives seemed to be Western Union's continued dominance of the telegraph industry or a government system integrated with the Post Office.

Western Union's Dominance as a Politico-Economic Problem

Postal telegraph advocates offered two main sets of reasons why the government should operate the nation's wires. On the negative side, they feared that Western Union posed a danger to the nation's economy and political institutions because it controlled the flow of news and financial information and exercised undue political influence. On the positive side, they hoped to transform telegraphy from an expensive medium used by a small percentage of the population into an affordable means of communication accessible to ordinary Americans.

Since about the mid-1850s, the fortunes of the newspaper press and telegraph industry had been tightly woven together. That connection intensified during and after the Civil War, resulting in a double-headed Western Union and NYAP monopoly over wire-service news. Editors who attempted to start rival newspapers found themselves frozen out of receiving NYAP dispatches. They had to resort to paying for dispatches at Western Union's standard commercial rates, forcing them to choose between paying high telegraph bills and forgoing national and international news. For example, when Henry George attempted to resurrect the *San Francisco Herald* in 1868, Western Union and the Associated Press prevented him from receiving dispatches because of complaints from the four established San Francisco dailies. George later claimed that this incident converted him into a postal telegraph advocate and launched his career as an economic reformer.[20]

Rumors swirled that Western Union had used its control over the nation's information flows to shape the political landscape to advance its own objectives. The contract between Western Union and several press associations stipulated that member newspapers were not to patronize rival telegraph lines or to publish stories or editorials detrimental to Western Union's interests. For example, in 1867 Western Union forced NYAP to stop supplying dispatches to the *Omaha Republican* because that paper had called the telegraph company a "grievous," "onerous," and "gigantic" "monopoly" in one of its editorials. In 1869 Orton thought that the New York newspapers that composed NYAP gave Western Union their "earnest support" during a key postal telegraph debate in Congress because the company had threatened to start a rival newsgathering service had they not done so. Thus, there was some justice to the charge that the newspaper press of the country stifled the views of postal telegraph advocates and advanced Western Union's instead. Reformers often attributed their failure to get Congress to pass a postal telegraph bill to this slanted press coverage. Similarly, Western Union's critics claimed that the company had used its influence over the press to favor Republican candidates Rutherford B. Hayes and James G. Blaine in the elections of 1876 and 1884. In the former case, critics charged that Western Union had done so because the Democratic nominee Samuel J. Tilden would have chosen Montgomery Blair, a well-known postal telegraph advocate, as his postmaster general. In the case of the 1884 election, many Democrats charged that Jay Gould had used his control over Western Union to spread the rumor that Blaine had won in order to manipulate the stock market for his own speculations.[21]

As this latter incident indicates, concern over Western Union's control of the flow of financial information intensified after Gould obtained hold of the company in 1881. Although there is no hard evidence that any officers or directors of Western Union manipulated the country's financial news, this suspicion had widespread currency, and not just among postal telegraph advocates. In a biography of his father-in-law John Pierpont Morgan, Herbert L. Satterlee recalled that Morgan acquired enough Western Union stock to become a director in 1882. By doing so, Morgan hoped to weaken Gould's influence over the company's affairs, because "news of every kind that was of importance in the stock market passed through the telegraph company's offices." Just after the turn of the century, muckraker Gustavus Myers similarly charged that Gould had routinely used his control over Western Union's wires to manipulate the stock market and to place news stories hostile to "any movement or agitation threatening the complete sway of capital."[22]

Advocates of a postal telegraph charged that Western Union not only controlled the press and the flow of financial news but also improperly influenced state and federal legislatures. Most especially, they criticized the company for liberally distributing franks, passes for free telegraph service, to public officials. Western Union managers publicly claimed that they gave out franks merely as courtesies. However, they privately admitted that franks were indeed a valuable investment, the "cheapest means" to stave off hostile legislation. On several occasions, Orton directed Washington, D.C., managers to overlook heavy telegraphing done by prominent members of Congress such as Schuyler Colfax and James G. Blaine. In 1869 Orton apologized to New Jersey senator John Potter Stockton for "unnecessary severity" on the part of office managers in limiting his free telegraphing, concluding that we "rely confidently upon your future efforts" to look after Western Union's interests. Other legislators, including James Garfield, Benjamin Butler, and Elihu Washburne, refused to accept franks, rightly regarding them as soft bribery. Reformers such as Frank Parsons frequently expressed outrage that the company engaged in "this insidious sort of bribery to influence legislation and administration in its own interests." At the very least, Parsons claimed, "A Congress that enjoys the privilege of free telegraphy will not be likely to vote down the system that gives them so valuable a privilege." Indeed, in 1891 the *Chicago Daily Tribune* claimed that the liberal distribution of franks was a key reason for the defeat of Postmaster General John Wanamaker's plan for a limited postal telegraph.[23]

It is difficult to determine exactly how much free telegraphing Western Union provided lawmakers, with estimates ranging from tens to hundreds of thousands of dollars annually. In 1891 the *Chicago Daily Tribune* estimated that seven-eighths of U.S. congressmen used Western Union franks and that they collectively sent about 150 messages a day while Congress was in session, some $1,000 worth of free telegraphing per month. However, public statements by company officers and internal records both indicate a higher figure. President William Orton told shareholders in 1873 that government officials accounted for nearly a third of the company's free business, with the rest being done by railroads in exchange for exclusive rights-of-way. In 1884 Western Union's vice-president John Van Horne testified before the Senate that the company provided about $1 million worth of free service. According to Orton's and Van Horne's figures, Western Union gave public officials about $300,000 in free telegraph service per year in the 1870s and 1880s. This is in rough agreement with internal company statistics showing that the company provided about $200,000 worth of "complimentary" telegraphing at the turn of the century.[24]

Some critics accused Western Union of going far beyond soft bribery, maintaining that the company perverted the political process itself. William McCabe of the International Typographers' Union claimed in 1894 that Western Union had used its political and financial clout to ensure the defeat of California congressman Charles A. Sumner, a strong postal telegraph advocate, when he ran for reelection in 1884. Frank Parsons charged that Jay Gould spent $250,000 to defeat Postmaster General Wanamaker's plan for a limited postal telegraph. He also understood "from inside information" that Gould had attempted to bribe Wanamaker with $1 million to withdraw his plan.[25]

Although Orton claimed that Western Union's efforts to gain political influence were "conducted with entire propriety," internal company records confirm some of McCabe's and Parsons's suspicions, if not their scope. In an era known for corruption and influence peddling, it is not surprising that Western Union attempted to shape the political landscape in some morally questionable ways. For example, in September 1869 Orton learned that Senator Alexander Ramsey of Minnesota, a leading postal telegraph advocate, owed much of his political success to a St. Paul newspaper. Orton considered buying Ramsey's silence by giving the *St. Paul Press* $10,000, in exchange for which the newspaper "would undertake to control the Senator and prevent him from pursuing a course opposed to our interests." Nothing came of this because Orton could not devise a method to get the money to the newspaper without Western Union's role being detected. Ramsey continued to push for a postal telegraph until he left the Senate in 1874.[26]

Western Union also maintained a major lobbying operation in Washington. During the late 1860s and early 1870s, the *New York Herald* reported extensively on the company's congressional lobbying activities, charging that James G. Blaine and John F. Farnsworth, the chairs of the House Appropriations and Post Offices and Post Roads Committees respectively, conducted themselves as if they were paid agents of the company. Indeed, Orton wrote privately that Blaine was a "warm" but "confidential" friend of the company. The *Herald* also discovered that the clerk of the House Post Office committee, Uriah Hunt Painter, was secretly on the company's payroll. Orton's letterbooks confirmed this, revealing that Painter distributed franks to congressmen, relayed sensitive information to Orton about pending legislation, and attempted to influence the votes of committee members on postal telegraph bills. Furthermore, Western Union presidents William Orton and Norvin Green routinely used their friendships with prominent politicians to influence legislation. Orton asked New York senator (and Western Union director) Edwin D. Morgan for help on several occasions, especially to revise the 1866 National Telegraph Act and to intercede with President Grant in 1872 to resolve a festering dispute between Western Union and the Army Signal Service. Similarly, in 1883 president Norvin Green asked fellow Kentuckian John G. Carlisle, who had just become Speaker of the House of Representatives, to select committee chairs favorable to Western Union's interests.[27]

In 1882 William H. Vanderbilt famously declared in a newspaper interview that he found "anti-monopolists most purchasable" among public officials: "They don't come so high." This cynical remark aptly described Western Union's relationship with William A. A. Carsey, active in the labor, agrarian, and greenback movements from the 1870s to the 1890s. Although he was not one of the major figures in these movements, he had earned enough respect to testify before Congress and to attend the 1890 Farmers' Alliance convention at Ocala, Florida, that gave birth to the Populist Party. In 1884 he described himself as an activist who sought to "regulate or abolish oppressive monopolies," and by 1888 he headed a small but vocal National Anti Monopoly League. Despite his background and views, he strongly opposed a postal telegraph. He told the Senate Committee on Interstate Commerce in 1888 that he instead preferred "moderate and discreet regulation," as long as it did not "injure the business." Carsey's opinions, in fact, were identical to Western Union president Norvin Green's. Green welcomed placing the telegraph under Interstate Commerce Commission jurisdiction because he thought it would kill off the postal telegraph movement for at least a decade and possibly for good. The reason for this remarkable congruence of views was that Carsey was on Western Union's payroll. Norvin Green thought that Carsey had "considerable influence" with reformers and legislators, and that he had performed "most valuable services in fighting antagonistic legislation at Washington." Green later paid his expenses to go to the Ocala Farmers' Alliance convention to work against a postal telegraph plank in its platform.[28]

While reformers feared Western Union's undue influence on the nation's news flows and political institutions, they also supported a government telegraph for a positive set of reasons. They hoped to transform the telegraph from an expensive business medium into a medium affordable and accessible to ordinary Americans. In 1883 perennial postal telegraph advocate Gardiner Hubbard framed the debate succinctly for the readers of the *North American Review*, "As a telegraph for business, where dispatch is essential and the price is of little account, the Western Union system is unrivaled; but as a telegraph for the people it is a signal failure."[29] Critics like Hubbard attributed this failure to the high rates the company charged in order to pay lavish dividends on inflated capital. A government telegraph, on the other hand, would merely need to defray construction and operating costs and could charge a fraction of Western Union's rates. Furthermore, reformers pointed to two successful examples of state intervention to make communications more affordable, the American cheap postage movement of the 1840s and 1850s and the nationalization of the British inland telegraphs in 1870.

Complaints about high rates initially sparked the postal telegraph movement. One of the major reasons why Senator B. Gratz Brown of Missouri introduced his pioneering postal telegraph bill in February 1866 was "telegraph extortion." Brown contended that the policy of the existing telegraph cartel was to charge "high rates for a limited business." A government network, however, would adopt the "post office principle," a uniform, low rate that would bring telegraphing within the reach of many more citizens. During floor debate on the National Telegraph Act later in 1866, Senator John Sherman characterized telegraphing as "a luxury for the rich." However, even the rich complained about high rates. Financier Jay Cooke repeatedly urged Sherman to continue his efforts at telegraph reform because "the public are now *fleeced*" by rates that were triple what they had been before the war. Even Western Union executive James D. Reid later admitted that the company's rates at the time reflected a "policy of rapacity."[30]

Telegraph managers responded to charges of high rates and extortionate behavior by claiming that their main customers—the press and businessmen—prized speed and reliability over cost. George Thurston, the president of the Pacific and Atlantic Telegraph Company, told Congress in 1873 that "no man sends a telegraph message for the mere pleasure of telegraphing. He merely sends it in order to gain time, or to make the profit which he expects to gain by economizing time, by forestalling the market . . . ; and the question with him is not so much the cost of a message as the speed with which it is delivered." James Brown, the head of the Franklin Telegraph Company was somewhat more flippant: "When a man pays 25 or 30 cents for a cigar, he does not care very much whether he pays 5 cents more or less for a telegraph message." William Orton of Western Union frequently noted that the company's true competitors were not rival lines but transmission and delivery times. Thus, telegraph company officials claimed, the telegraph was by its very nature a business medium whose customers willingly paid well for a premium service.[31]

TABLE 2.1.

Western Union's Par Capitalization, 1851 to 1893 (in thousands of dollars)

Date	Capitalization	Notes
1851–59	370	
1859	2,300	Several issues of stock dividend
Dec. 1863	11,000	Several issues of stock dividends
May 1864	22,000	100% stock dividend.
1866	41,000	Buyouts of United States and American Telegraph companies
1879	41,000	"Cutting the Melon": 36% dividend, including $9.2 million in cash and $6 million in stock held in company treasury
1881	80,000	$15 million and $8.4 million issued respectively for purchase of American Union and Atlantic & Pacific Telegraph companies, and $15.5 million stock dividend to shareholders owning stock before these acquisitions
1888	86,200	Acquisition of Baltimore and Ohio Telegraph Company.
1893	94,820	$8.62 million stock dividend

SOURCES: Thompson, *Wiring a Continent*, 396–426; Reid, *Telegraph in America*, 484–85; WUTC Annual Reports.

According to postal telegraph advocates, however, high telegraph rates were not due to the costs of providing an exacting clientele with first-rate service. They arose instead from Western Union's need to pay large dividends on its inflated capital. Indeed, from 1859 to 1866, Western Union's capitalization swelled a hundredfold, from about $370,000 to $41 million. Although the company had grown from a regional telegraph line operating in the Great Lakes region to a national telegraph network in these years, its expansion could not account for all of this vast increase in capitalization. Indeed, Western Union's capitalization was roughly double the value of its physical plant and triple the market value of its stock for several years after the war. In 1868 the company's shares traded at about a third of par value, and in 1869 William Orton privately admitted that the true value of Western Union's property was about $25 million.[32]

Orton's overriding priorities were to stave off a government telegraph and to make private competition difficult if not impossible. He therefore focused on Western Union's long-term prospects, not on the current stock price or size of quarterly dividends, as his major stockholders might have wished. When he assumed control of the company in 1866, he began to wring out the water in the company's capital while gradually reducing rates. Between 1866 and his death in 1878, Western Union nearly tripled the mileage of its lines and number of offices and quadrupled the number of messages handled. Orton slashed dividends to fund this construction and to bring some $7.3 million of stock into the company's treasury, effectively reducing its capitalization to about $34 million. At the same time, he reduced the cost of the average telegram from $1.05 to $0.39. By doing so he weakened much of the antimonopoly sentiment that served as a rationale for a postal telegraph and a rallying cry for would-be competitors. Even Western Union's strongest critics—for example, Francis B. Thurber of the New York Board of Trade and Transportation and postal telegraph advocate Gardiner Hubbard—praised Orton's efforts to reduce rates and to align the company's capitalization with its book value.[33]

A few months after Orton's death in April 1878, the company's prominent shareholders revolted. Led by William H. Vanderbilt, they demanded a combined cash and stock dividend of $17 million, equal to about half the market value of the company's outstanding shares. The dividend paid out in 1879 was just a bit short of this figure, amounting to $9.2 million in cash and $6 million of the shares held in the company's treasury. After Jay Gould acquired the company in 1881, the influence of shareholders demanding immediate returns increased. That year Western Union nearly doubled its capitalization to $80 million, issuing some $23.4 million shares to cover the purchase of the Gould-controlled Atlantic and Pacific and American Union Telegraph companies and rewarding its existing shareholders with a stock dividend of $15.5 million, nearly 40 percent. Orton's successor, Norvin Green, also reversed the practice of paying dividends only after the construction and repair accounts had been funded. Henceforth the company distributed nearly all its profits as dividends and covered construction and repair costs through dividend income from the stocks of other companies held in its treasury, including large stakes in the fast-growing ticker and telephone businesses. To place the issue of capitalization and dividends in another light, between 1866 and 1885 Western Union paid shareholders a total of some $44 million in cash dividends, paying out over half of this total amount, some $23 million, after Gould had acquired control of the company in early 1881.[34]

Thus, Vanderbilt, Gould, and Green adopted a more modern yardstick for evaluating Western Union's worth. Orton had hewed to the traditional idea that the company's capitalization should reflect the value of its physical plant. Green and prominent investors argued instead that the company's true value lay in its ability to earn profits, not the cost of its wires and equipment. For instance, in his 1883 testimony before the Senate Committee on Relations between Labor and Capital, Green insisted that the company was worth well over its nominal capitalization, amounting to perhaps $100 million, if one included not only the physical plant but also such intangibles as its patent portfolio, railroad rights-of-way, goodwill, and most especially its future earning capacity.

To a certain extent Western Union's high capitalization was a public-relations move. Despite persistent criticism of the company's nominal capital, Green thought that paying moderate dividends on a large capital was "less likely to excite antagonism as an extortionate monopoly" than paying high dividends on a smaller capitalization that matched the company's book value. Similarly, Orton and Green repeatedly claimed that Western Union was not a corporation closely held by wealthy New York capitalists, insisting instead that several thousand people owned shares. These investors of modest means, including the proverbial widows and orphans, deserved a fair and moderate return on their investments. Finally, a higher capitalization made it less likely that Congress would buy the company out under the 1866 National Telegraph Act. Indeed, many postal telegraph advocates opposed a buyout of Western Union's lines, because, as Senator Shelby Cullom phrased it in 1888, "they cannot be acquired at anything like a fair valuation."[35]

Postal telegraph advocates remained skeptical of Orton's and Green's claims about Western Union's capitalization. In fact, their claims gave renewed force to the postal telegraph movement. For example, Colorado senator Nathaniel P. Hill introduced a postal telegraph bill in 1884, one provision of which called for the Senate to investigate "whether the cost of telegraphic correspondence . . . has been injuriously affected by large stock dividends made by the Western Union Telegraph Company." In his accompanying report, Hill claimed that five-sixths of the company's $80 million capitalization was water and that this "swollen capitalization" was "a cover, an inducement, and in some senses a necessity for excessive charges for telegrams."[36]

Reformers like Hill insisted that a postal telegraph, freed of the necessity of paying out huge dividends, would operate under the same public-service ethos that the mails did. The high standard of service set by the postal system was thus a major positive argument for nationalizing the telegraph. Virtually all Americans—even diehard opponents of the postal telegraph—regarded affordable and convenient mail delivery as a basic republican right. An important strategy of reformers, therefore, was to demonstrate the essential identity of the telegraph and the mail. They argued that the government monopoly on mail delivery also covered telegraphy: had the telegraph been in existence in the 1780s, the framers of the Constitution would have surely included it among the postal department's duties. Postal telegraph advocates drew particular inspiration from the cheap postage movement of the 1840s and 1850s, which transformed the mail from largely a business medium into a social medium. After 1851, when Congress lowered the cost of sending a letter anywhere in the country to three cents, social and family correspondence increased dramatically. Telegraph reformers wished to reduce rates and to extend facilities in order to transform the telegraph into a social medium, just as their antebellum forebears had done for the postal system.[37]

The nationalization of the telegraph in Britain was also an inspiration to American reformers. The British Post Office began operating the inland telegraphs in February 1870, and the service initially fulfilled reformers' optimistic expectations. During its first five years of operation, Post Office administration of the British telegraphs was indeed impressive: the average cost of a telegram dropped by nearly half, the number of messages more than tripled, the number of offices nearly tripled, and press dispatches increased tenfold.[38] American advocates of a postal telegraph hailed these gains as sure signs of the success of British postal operation and of the comparative failure of Western Union to provide affordable and accessible service. For them, the British example furnished a ready-made model of the social benefits of government ownership.[39]

Looking to the examples of the cheap postage movement and the British postal telegraph, reformers continually claimed that an American government telegraph would slash rates and dramatically increase the number of messages handled, while still breaking even. For example, in 1869 Congressman Elihu Washburne and Gardiner Hubbard both claimed that a postal telegraph could pay expenses at rates of ten cents for ten words for distances up to five hundred miles, with an additional five cents for each additional five hundred-mile distance (not including delivery charges). By contrast, Western Union's average toll per message was seventy-five cents in 1870. Reformers claimed that these huge cost savings would come from three sources. They made an analogy between cheap telegraphing and the cheap postage movement of a generation before. Three-cent postage, after all, had little effect on postal expenses despite the much higher volume of letters handled. A government telegraph need not pay dividends on inflated capital but would only need to defray construction and operating costs. And a government telegraph could take advantage of certain economies, such as placing telegraph offices within post offices, using letterboxes and mail carriers to deliver messages, and training postmasters as telegraphers.[40]

Reformers' claims that a government telegraph could charge such low rates while breaking even seems unlikely. Consider first the British experience after the 1870 nationalization. During the 1860s, British advocates of nationalization argued from analogy to the penny post that lower rates would result in a vast increase in message traffic and that income and expenses would both remain fairly constant as the volume of telegrams increased. However, the British telegraphs lost more and more money throughout the rest of the century, a total of some

$25 million by 1895. The increasing losses of the British postal telegraph served as a powerful deterrent to an American postal telegraph as time went on.[41]

A major reason for the chronic British deficit was the radically different marginal cost structures of the mails and telegraphs. Two hundred pieces of mail could be transported and delivered for much less than double the cost of transporting and delivering one hundred, a marginal cost structure that made cheap postage financially practicable. However, each telegram required two operators to send and receive it and monopolized the circuit during its transmission. Telegraphy thus could not take advantage of the same economies of scale that the mails did. For example, labor costs, about 60 percent of the costs of handling telegrams, remained constant per message no matter how much traffic increased.

Furthermore, a postal telegraph was likely to increase labor costs. Western Union typically used low-paid messenger boys to deliver telegrams, whereas a postal system would have replaced them with adults, comparable to its letter carriers. Also, an important reason why reformers, particularly those from the labor movement, advocated for a government telegraph was to improve operators' wages and hours. Finally, the British experience showed that extension of civil service protections to telegraph operators resulted in reduced worker effort. British telegraph engineers Henry Fischer and William Preece noted in 1877 that a typical British Post Office operator handled about half the message volume that a Western Union operator did owing to civil service protections.[42]

Finally, reformers' claims that Western Union's high capitalization was responsible for its high telegraph rates were also questionable. Western Union's annual reports show that its average cost per message ranged from about 65 to 75 percent of its average toll from 1866 to the turn of the century. Eliminating the need to pay dividends to shareholders would thus lower rates by only that approximate percentage. For these reasons, it seems unlikely that a government telegraph in the United States could charge significantly lower rates than Western Union without running annual deficits. Indeed, government control of the wires in 1918 and 1919 would later bear this out. During the year of government control rates in fact rose 20 percent.

However, postal telegraph advocates believed that the social benefits of a government telegraph far outweighed any potential drain on the treasury. Visiting the 1893 World's Columbian Exposition in Chicago, British telegraph engineer William Preece declared that the British telegraph was more "republican" than its American counterpart because it "belonged to the people." He acknowledged that the British postal telegraph lost about $2 million a year, but he believed that the social and educational benefits were "worth this additional charge on the taxes." In 1900 Frank Parsons echoed Preece, arguing that the "civilizing and social benefits" of a government telegraph were "beyond computation in money."[43] Preece's and Parson's remarks suggest that reformers saw a vastly different social and economic role for the telegraph than did Western Union officials. Whereas telegraph managers catered primarily to speculators, businesspeople, and press associations, reformers like Parsons wanted it to fulfill a public service ideal. For them, the telegraph ought to provide affordable and convenient service to the people at large and to educate the public by disseminating news as widely as possible.[44]

Options to Address Western Union's Dominance

To contend with Western Union's dominance of the telegraph industry, postal telegraph advocates proposed a bewildering array of options between 1866 and 1900, ranging from less to more government intrusion into the industry. The least intrusive options aimed to use the power of the federal government to encourage telegraph competition, either by using the Army Signal Service's weather reporting network to patronize rival lines or by passing legislation permitting or requiring railroads to operate public telegraph services. A more active option was to have the Post Office Department contract with private telegraph companies to connect post offices around the country with a network of leased wires. This was analogous to contracts that the Post Office routinely signed with railroads, steamboat lines, and stagecoach companies to transport the mails, as well as to various proposals for a parcel post to give citizens a low-cost alternative to the express companies. A third set of options proposed building a government telegraph to compete with Western Union. These plans varied in scale and expense from a short line on the Eastern Seaboard to a comprehensive national network. Finally, many reformers proposed buying out the existing telegraph companies entirely. The inspiration for this was the nationalization of the British inland telegraphs in 1870.

The consolidations of 1866 sparked the fifty-year agitation for a government telegraph. By the end of July 1866, Congress had considered three alternatives. The first was to set up a government network for postal purposes that would operate alongside the privately owned telegraph companies. Another was to grant new companies special privileges to allow them to compete effectively with Western Union. Neither of these alternatives passed Congress in 1866. A third did—the National Telegraph Act, which empowered Congress to buy out private telegraph companies after 1871.

In February 1866 Missouri senator B. Gratz Brown succeeded in passing a resolution directing Postmaster General William Dennison Jr. to report on the feasibility of constructing a government telegraph network along the major railroad routes that carried the mails. Brown contended that the policy of the existing telegraph cartel was to charge "high rates for a limited business." A government network, however, would adopt the "post office principle," a single uniform and low rate that would bring telegraphing within the reach of many more citizens. More seriously, Brown claimed that NYAP, with the cooperation of the telegraph companies, had established a dangerous monopoly over the flow of telegraphic news. A government telegraph would supply the news to any newspaper desiring it "at a trifling cost and with infinite advantage to the nation."[45]

Dennison reported back at the beginning of June 1866, recommending against Brown's project because of its "doubtful financial success" and "questionable feasibility under our political system." Advocates and opponents of a government telegraph would echo Brown's and Dennison's respective arguments for the next thirty years. Postal telegraph supporters would continue to claim that a government network would place telegraphing within reach of the ordinary citizen and that it would break up the Associated Press's monopoly of the news. Skeptics would counter that a postal telegraph was too costly an undertaking, that it would increase the executive branch's patronage army, and that government ownership ran counter to the spirit of republican institutions.[46]

Dennison's report was far from impartial. It relied on a reply from American Telegraph Company electrician George B. Prescott and a joint letter from the presidents of the three commercial lines, Edward Sanford of the American, Jeptha Wade of the Western Union, and William Orton of the United States Telegraph companies. Their replies merit discussion because they revealed the contours of a long debate over the nature of monopoly. Prescott agreed with Senator Brown that there was a "striking analogy" between telegraphy and the mails, because "one company can manage the service better than two or more." The three presidents similarly claimed that competition in telegraphy "cannot be maintained." Yet they also denied that the consolidation of the three companies was a monopoly in the strictest sense, based on the acquisition of special privileges, but that the consolidation had occurred instead because of "the pressure of mutual necessity." Despite Western Union's dominance of the industry after 1866, its managers would continue to claim that it was not a monopoly in this narrow sense because it had received no special privileges from state or federal governments. After all, anyone was free to obtain a state charter and incorporate a new telegraph company. Reformers responded that a natural monopoly was just as dangerous and insulated from competition as a fiat monopoly.

Prescott and the company presidents also contended that a government telegraph would be prohibitively expensive. Prescott estimated the cost of a national network at some $45 million, basing this on a construction cost of more than $400 per mile of single-wire line, nearly triple his estimated cost of $150 per mile that he gave in his 1860 telegraph engineering treatise. Similarly, the three presidents estimated the cost of a six-wire line from Washington to Boston at $250,000, or about $660 per mile. Although they blamed wartime inflation and the increased cost of improved wires and insulators as the reasons for this high figure, critics of Western Union charged that the company's claims of inflated construction costs really reflected the huge amount of water in its capitalization.

A few days after Brown introduced his bill, the Senate debated another important telegraph measure. The International Ocean Telegraph Company, a New York corporation set up in 1865 to build cables to Cuba and the West Indies, sought legislation granting it an exclusive landing license for twenty-five years. In other words, this bill would have granted it a twenty-five-year monopoly over all telegraphic communication between the United States and the Caribbean islands. Several senators objected to this exclusive landing license, particularly Ohio's John Sherman. They sought to make the bill's terms general, to grant landing privileges to any company that built a cable from the United States to the West Indies. Debate went beyond the desirability of a cable to Cuba and became instead a general discussion of the merits of government telegraphy. Sherman, for example, hoped for "an entire revolution in the system of telegraphing. As it is now, it is only a luxury for the rich." He thus "heartily sympathize[d]" with Brown's concurrent proposal to set up a government telegraph "to break down the . . . monopoly of the telegraph companies."[47]

Sherman's remarks were consistent with his historical reputation as a foe of trusts and monopolies. However, in early April he introduced a bill to grant a newly formed company a congressional charter to compete with Western Union. As Sherman originally drafted it, this bill would have given valuable and exclusive privileges to this so-called National Telegraph Company, particularly the right to build lines along any post road, including any railroad line used to carry the mails. Sherman thought that this privilege would break down Western Union's main competitive advantage, its contracts with the major railroads for exclusive rights-of-way. Several senators wanted to make the measure general, so that any telegraph company agreeing to these terms could take advantage of the bill's privileges. Sherman resisted this, claiming that the only way to ensure real competition with the new Western Union monopoly was to back the National Telegraph Company. However, Sherman was hardly disinterested. The incorporators and investors of the National Telegraph Company included several of his friends and relatives, including his brother Charles and brother-in-law Thomas Ewing. Critics of Sherman's measure noted correctly that the company was simply a stock-jobbing scheme. During floor debate, the Senate thoroughly amended the bill to make its provisions general.

Sherman himself was ambivalent about the bill after its final passage. Although his personal papers during the first half of 1866 make frequent mention of the National Telegraph Company and the bill he authored to grant it a federal charter, his memoirs and biographies do not. During the final debate on the bill, he even disavowed responsibility for its amended version, leading Nevada senator James Nye to exclaim that the bill "is a child of your begetting, and I am a little surprised that the Senator should disown it." Perhaps Sherman did not care to have his name associated with a measure that smacked of nepotism and influence peddling.[48]

The final bill, the National Telegraph Act, contained three major provisions. Its framers intended it to foster competition by giving any telegraph company agreeing to its provisions the right to build lines on any road that Congress had declared a post road, especially railroad lines that carried the mails. In exchange for this privilege, telegraph companies agreed to two further stipulations, both powerful tools to regulate the telegraph industry. The law allowed the postmaster general to set the rates that a company could charge for government business. The act also gave Congress the option to buy out any company that accepted the bill's provisions after five years had elapsed.

The National Telegraph Act of 1866 was the only piece of legislation giving the federal government oversight of the telegraph industry until the Mann-Elkins Act placed it under Interstate Commerce Commission jurisdiction in 1910. Although the 1866 act did not regulate telegraphy directly, taken as a whole it created a quasi-regulatory environment that encouraged competition and restrained Western Union from using the full force of its monopoly power. The provision granting the postmaster general authority to set rates for government messages became important after 1870, when the army's Signal Service began a weather reporting system that used the national telegraph network extensively. The second provision, that after 1871 Congress could buy out any telegraph company agreeing to the act's terms, was a strong inducement for Western Union to reduce its rates and improve its service in order to blunt antimonopoly sentiment that might lead to such a buyout.[49]

When the National Telegraph Act became law in July 1866, Western Union's managers hesitated to sign it and to expose the company to the act's two quasi-regulatory provisions. Yet in June 1867 the company signed. It did so to protect its extensive rights-of-way along railroads in the South. After the war, many southern railroads were reorganized under new managers, and Western Union had difficulty getting them to recognize the telegraph company's rights-of-way. Also, many states treated foreign and domestic corporations differently, placing Western Union at a disadvantage when negotiating with railroads or undertaking legal action to assert its rights. Western Union executives therefore welcomed a measure of federal regulation to transcend the patchwork of state laws and court systems. Because many southern states fell under military occupation, Western Union also sought military protection against any interference with its property rights.[50]

Signing the act seemed to carry little risk to Western Union. Until the army set up a national weather reporting service in 1870, government telegraph traffic was light, so in 1867 Western Union managers placed little importance on the provision of the act granting the postmaster general authority to set rates for government messages. Similarly, William Orton and Norvin Green both thought that prospects for a government buyout after 1871 were slim. Orton was confident that by 1871 he could demonstrate that Western Union provided "a service so satisfactory to the public that thereafter there would be no demand from any quarter for intervention by the Government."[51] Even if Congress attempted to exercise its buyout option in the future, Western Union could hold out for a high enough sale price to satisfy shareholders or to discourage Congress from proceeding.

During the late 1860s and early 1870s, three concurrent developments structured the postal telegraph debate. In 1870 the army's Signal Corps launched a national weather-reporting service. Western Union engaged in a long-running dispute with the War Department over rates and terms of service, going so far as to refuse to handle these messages for several months. Western Union's intransigence provided yet another indication to its critics that the company was a soulless monopoly that cared little about serving the public. Also, the final 1874 agreement between Western Union and the War Department froze competitors out of handling this business, stifling telegraphic competition and further cementing Western Union's dominance. This episode was yet another demonstration of the futility of telegraphic competition and helped convince many Americans that the government should operate a telegraph system if only for its own purposes. At the same time, reformers led by Wisconsin congressman Cadwallader Washburn and President Grant's postmaster general John A. J. Creswell sought to nationalize the private telegraph industry along the lines of the National Telegraph Act and the 1870 nationalization of the British telegraphs. But outright nationalization had two main problems, even for some of Western Union's strongest critics: the government would be forced to pay an inflated price for Western Union's lines, and it would result in a huge increase of the federal government's patronage work force. Boston businessman Gardiner G. Hubbard offered an alternative that promised cheap telegraphing without requiring the federal government to own and operate telegraph lines. Hubbard tirelessly promoted a hybrid plan to incorporate a United States Postal Telegraph Company (not to be confused with the Postal Telegraph Company that competed with Western Union from 1883 to 1943) that would contract with the Post Office to transmit cheap telegrams in exchange for a guaranteed return of 10 percent on its capital. Despite the manifest self-interest of Hubbard's scheme, it proved to be the most durable and long-lived of all the postal telegraph proposals introduced between 1866 and 1890.

The use of the telegraph to gather weather reports and issue predictions had begun in 1849 when the Smithsonian's secretary Joseph Henry asked telegraph companies for free use of their lines for this purpose. By the Civil War, Henry collected and displayed daily observations from about five hundred reporting stations around the country. The Civil War and a devastating 1865 fire at the Smithsonian discontinued the weather reports, but in 1869 scientists Increase Lapham of Milwaukee and Cleveland Abbe of Cincinnati began agitating for their renewal. A year later, Brigadier General Albert Myer, head of the army's Signal Corps, took up the cause mainly to provide a rationale for the continued existence and expansion of the Signal Corps. By 1869 the corps staff had shrunk to four, including Myer, and its budget to a minuscule $5,000. The army would continue to operate it until 1891 when the Department of Agriculture took it over and the Signal Corps returned to purely military duties. By all accounts, the army's weather reporting service was a resounding success. Storm warnings on the Great Lakes were about 70 percent accurate and they annually saved more than a million dollars in property that would otherwise have been lost in shipwrecks.[52]

Because the Signal Service did not operate its own telegraph network for weather reporting purposes,[53] it depended on the lines of the private companies to gather weather observations and disseminate local predictions. Although Myer initially praised telegraph company managers for their "liberality and fairness," Western Union and the Signal Service endured a fraught relationship between 1870 and 1874. Western Union refused to handle Signal Service messages from March to May 1871 and again from July 1872 through the beginning of 1874. Western Union officials complained that the rate of compensation was too low, that the work was too exacting, and that weather reports took over important circuits during peak business hours. There was some justice to Western Union's complaints. Although it is impossible to verify the company's claims that it did the work at a loss, the rates were far lower than for comparable private message traffic. Also, Myer used a complicated cipher to relay weather observations and a complex transmission plan that rerouted many of Western Union's circuits during the three times daily that Signal Service observers sent their reports to Washington. But the real issue was that Western Union insisted on treating the Signal Service as merely another customer. Similar to its contracts with press associations, Western Union demanded a monopoly of all government business. The company also required prepayment of all messages, refusing to allow the Signal Service to settle accounts monthly.[54]

Orton apparently thought that the government would accede to his terms, since Western Union was the only telegraph company capable of providing nationwide service. However, Myer turned to a coalition of Western Union's rivals—the Franklin, Atlantic and Pacific, Pacific and Atlantic, and Southern and Atlantic Telegraph companies. These companies were eager to handle the weather reports, and the Signal Service was able to continue its reports with minimal disruption during the two Western Union blackouts. The Signal Service encouraged the independent companies to cooperate with each other and to extend their lines, leading to hopes that they might form a united opposition. Indeed, for a time in 1872 and 1873, Signal Service business proved important for these companies to maintain themselves as competitors to Western Union.[55]

Western Union's conflict with the Signal Service helped shape the contours of the postal telegraph debate. Orton took a hard line with Myer because he wanted to challenge the legality of a key part of the National Telegraph Act of 1866. Orton conceded that the act gave the postmaster general the authority to set rates for government messages but protested that it did not give Myer the authority to demand special circuit makeups or priority of transmission. In response, the War Department obtained legal opinions from the U.S. attorney general's office asserting that government departments had wide-ranging authority over the manner in which private telegraph companies handled their messages. A more important reason for Orton's hostility was that he believed that Myer harbored a long-standing "ambition to establish a Government system of Telegraph, of which he is to be the head."[56]

Postal telegraph supporters took Western Union's intransigence as further evidence of the need for reform. Iowa congressman Frank W. Palmer argued in February 1872 that Western Union's refusal to provide this important public service justified passage of Gardiner Hubbard's postal telegraph plan. Hubbard himself thought that Western Union's behavior would all but guarantee the bill's passage. Although Hubbard's optimism proved wrong, the controversy won new converts to a government telegraph. For example, Congressman James A. Garfield opposed the Hubbard plan as well as outright nationalization of the wires, but he came away from the weather service controversy convinced that the government ought to operate its own network for both postal and military purposes. Similarly, Western Union's refusal to handle weather messages was an important rationale for legislation in 1873 and 1874 authorizing the army to build its own telegraph lines to connect forts in the West and lighthouses and life-saving stations on the Atlantic coast. By the turn of the century, the army built some eight thousand miles of telegraph line and its operating mileage peaked at about five thousand miles in the early 1880s. Some of these lines did a light public business, but the army stopped handling civilian traffic as soon as commercial telegraph companies could serve these areas. Yet Orton strenuously opposed the construction of these lines, fearing that they might become the nucleus of a competing government network.[57]

In early 1874 the Signal Service threw over the independent telegraph companies and gave its business back to Western Union. The proximate cause was Western Union's absorption of the Pacific and Atlantic company, an important part of the independent coalition. The deeper reason, however, was that the independents could not provide the Signal Service with the same service and geographic coverage that Western Union had. Operators on the rival lines often had trouble mastering the War Department's weather cipher, resulting in mistakes in dispatches. The companies sometimes failed to forward messages over each others' lines and to connect their wires into the complicated circuit schedules that Myer required, delaying the transmission of observers' reports to Washington. The officers of the Atlantic and Pacific and Southern and Atlantic companies vigorously protested the transfer of the service back to Western Union. At the very least, they contended, Western Union's behavior should have convinced Myer of the wisdom of fostering telegraphic competition. Although Western Union frequently claimed that it did the Signal Service work at a loss, its rivals thought that Signal Service business had been moderately profitable and had given them a base traffic load that had helped them stay solvent. The Atlantic and Pacific in particular placed great importance on Signal Service traffic. An important reason for that company's acquisition of the Franklin in early 1873 was the large number of weather reports that it handled. The company tried to reverse Myer's decision several times between 1874 and 1877, offering to do the work for 10 percent less than Western Union charged. The loss of Signal Service business hit the Southern and Atlantic quite hard and forced it into a pooling agreement with Western

Union in the summer of 1874. The defection of the Southern and Atlantic deprived the Atlantic and Pacific of an important ally that provided it with valuable southern connections. The Atlantic and Pacific in turn entered a pooling agreement with Western Union in 1877.[58]

The Signal Service's retreat was thus a clear victory for Western Union. Although the 1874 settlement resulted in smoother relations and better handling of weather reports, it also stifled telegraphic competition and strengthened Western Union's dominance. Western Union insisted on an exclusive contract with the Signal Service, thus barring future competitors like the Baltimore and Ohio and American Union Telegraph Companies from doing a share of the business.[59]

Western Union's intransigence over handling Signal Service traffic and the failure of telegraphic competition reinforced reformers' belief that only a government-operated telegraph could solve the problem of Western Union's dominance. The first Grant administration was the high water mark of the postal telegraph movement, perhaps the closest it came to succeeding. Not only were Grant, Postmaster General Creswell, and prominent legislators enthusiastic supporters, but Western Union's managers and major stockholders seemed to accept the inevitability of a government telegraph. Despite their public opposition, they privately thought that they could sell out to the government or work within Gardiner Hubbard's hybrid plan on favorable terms. In Congress, the brothers Elihu B. Washburne of Illinois and Cadwallader C. Washburn of Wisconsin led the movement. They made postal telegraphy their leading (if not sole) priority during their last terms in the House of Representatives in 1869 and 1870. However, the brothers disagreed on the best way to provide the public with affordable telegraphing. Elihu Washburne introduced a bill in 1868 to build a competing government network for postal purposes and later supported Gardiner Hubbard's hybrid scheme. His brother, on the other hand, rejected both alternatives and called for outright nationalization on the British model, using the method specified in the National Telegraph Act of 1866.[60]

The inability of the Washburn(e) brothers to agree on a particular plan was symptomatic of a major and chronic reason for the failure of the postal telegraph movement. Despite widespread support for a government telegraph, reformers were unable to unite behind a specific measure. In January 1870 Cadwallader Washburn succeeded in creating a House Select Committee on Postal Telegraph Lines that sat throughout most of the 41st Congress. In July 1870 the committee reported two bills to the House: Washburn's, to buy the existing telegraph companies according to the provisions of the 1866 act; and another by Frank Palmer of Iowa, to incorporate Gardiner Hubbard's Postal Telegraph Company.

Committee members could not agree to back either plan, so they merely reported both and urged representatives to study them during recess. When Congress reconvened in December, the Select Committee remained deadlocked. Two of the seven committee members favored Hubbard's plan and only Washburn favored outright nationalization. James Beck of Kentucky was unalterably opposed to either bill. Thus, neither plan got the support of a majority of the committee. This remained the pattern in Congress over the next few years. Postal telegraph advocates in the House tended to favor nationalization per the 1866 act, but those in the Senate tended to back Hubbard's proposal.[61]

Many telegraph reformers could not go so far as to support outright nationalization for fear of increasing the federal government's patronage army and intruding too deeply into the private sector. Hubbard's plan appealed to them because it promised to provide cheap telegraphing without those two objections. It would remain a mainstay of the postal telegraph movement into the 1890s. Although Hubbard continually insisted that his company could achieve its targeted 10 percent rate of return with no government assistance, skeptics thought it likely that Congress would have to subsidize Hubbard's company to do so. In his January 1870 report, for example, Washburn criticized Hubbard's hybrid postal telegraph plan as a scheme to enrich himself and his backers at the taxpayers' expense. Although Hubbard proved to be a man of integrity even to his sternest critics, this common and persistent charge had some justification. As his private correspondence showed, Hubbard believed that he could do well by doing good—that he could make a tidy profit while giving the public cheap telegraphing.

Despite the inability of postal telegraph advocates to rally around a particular plan, the eventual postalization of the telegraph had an air of inevitability at the time. In January 1871 William Orton wrote privately that he expected a postal telegraph in the near future and thought that Western Union could give up its public message business while still profiting handsomely from its ticker and private-wire businesses. Other industry insiders claimed that the Vanderbilt interests that controlled the company since late 1870 welcomed nationalization per the 1866 act, because it would net them a neat return on their investment. Creswell's personal papers contain an unsigned memorandum dated December 1870 on Western Union letterhead offering to sell out to the government for $30 million. A year later, Congressman James Beck (a confidant of William Orton) claimed that Western Union was "most anxious" to sell out under the 1866 National Telegraph Act for the likely sum of $35 million. Also, Orton told Washburn's Select Committee in the spring of 1870 that Western Union would agree to handle public messages for the Post Office on the same terms that Hubbard sought, a 10 percent return on capital in exchange for a uniform telegram rate of twenty cents. In late 1870 and early 1871 Orton met privately with Hubbard to discuss how Western Union could work within his plan, and even invited him to explain it to Western Union's Executive Committee. Similarly, in late 1872 James Garfield, chair of the House Appropriations Committee, regarded the passage of a postal telegraph bill in the next (43rd) Congress as a foregone conclusion because of the Republicans' dominance of Congress, though he "greatly dread[ed]" its "centralizing" tendencies.[62]

Despite Garfield's expectations, the 43rd Congress (1873–75) did not pass a postal telegraph bill, and the subject did not even come up during the 44th (1875–77). Even Grant and Creswell stopped calling for nationalization in their annual messages to Congress after 1873. The main reason for this lack of activity was the renewal of telegraphic competition. The Atlantic and Pacific Telegraph Company was a formidable competitor from the early 1870s until it entered a pooling agreement with Western Union in 1877. After 1874 Jay Gould controlled the Atlantic and Pacific, and he used his vast railroad empire to build a national network to contend with Western Union. Although it would later become clear that Gould was using the Atlantic and Pacific as a cat's-paw to gain control of Western Union and reestablish a monopoly, telegraph customers welcomed the reduced rates that competition fostered.

The only significant piece of telegraph legislation that Congress considered between the early 1870s and early 1880s was an 1879 rider attached to an army appropriations bill by Massachusetts congressman Benjamin F. Butler. On the surface, this provision fostered telegraphic competition by permitting any railroad company that agreed to the provisions of the 1866 National Telegraph Act to conduct a public telegraph business. In reality, this provision served nobody's interests except Jay Gould's. A few months after its passage, Gould formed the American Union Telegraph Company to compete with Western Union. The Butler Amendment had the effect of negating the advantages of Western Union's exclusive agreements with railroads. Thanks in large measure to the Butler Amendment, Gould built the American Union into a formidable competitor of national scope. Competition from the American Union depressed Western Union's stock price, allowing Gould to buy enough shares to gain control in January 1881. This ended meaningful telegraphic competition for several years.[63]

Because Gould was one of the most vilified men in America, his control of the telegraph industry reanimated the postal telegraph movement. The period between 1881 and 1885 was thus another high water mark of the movement. Although prospects for passage of postal telegraph legislation were not as good as they had been in the early 1870s, Gould's control of Western Union convinced many skeptics of the dangers of a private telegraph monopoly. For example, in 1874 the influential engineering journal *Manufacturer and Builder* had strenuously opposed a postal telegraph and had even advocated for the privatization of the mails. In February 1881, however, the journal reversed itself, calling for nationalization on the British model because of Gould's takeover of Western Union.[64]

Immediately after Gould's takeover, the postal telegraph again became a leading public issue. Several congressmen introduced postal telegraph legislation in the House in January 1881, but none of these measures came to the floor because of an already crowded legislative calendar. The nation's prominent boards of trade began calling for telegraph reform for the first time as well, including the National Board of Trade and the New York Board of Trade and Transportation. The New York Produce, Cotton, and Petroleum Exchanges issued a joint circular noting their "grave suspicion" of Gould's renewed telegraph monopoly and calling for the establishment of a competing telegraph company controlled by the nation's commodity exchanges. The Chicago Board of Trade heartily endorsed this plan, "unless a system of Government Telegraphy can be adopted."[65]

Antimonopoly agitation reached a crescendo in the summer of 1883 in the wake of a bitter telegraphers' strike. Operators across the country had long protested sweeping wage reductions and work speedups, to no avail. At the same time, Western Union had declared some $18 million in dividends since Gould's takeover two years ago. These facts, along with the strikers' temperate conduct, ranged the public squarely with the operators. Although the company crushed the month-long strike, it was a public-relations disaster that caused the nation's business classes and newspapers to swing in favor of a postal telegraph. Stock and commodity brokers felt particularly aggrieved, because they saw their speculative business fall off dramatically during the strike. The American Chamber of Commerce, for instance, told the Chicago Board of Trade that the newspaper press had been "either apathetic or antagonistic" to postalization before the strike but had come "to favor the measure very decidedly" afterward. Christian reformer and editor Lyman Abbott made the same observation as well. Indeed, seven years later, Postmaster General John Wanamaker included some forty pages of newspaper editorials about the strike as the centerpiece of his report calling for a postal telegraph. The telegraphers' strike was also the most important catalyst behind the Senate's celebrated 1883 hearings on "the relations between labor and capital" that took up four volumes of testimony. The Senate's Committee on Education and Labor convened in New York that summer and fall and summoned

Norvin Green and Jay Gould to justify their actions during the strike. A few days after Western Union had crushed the strike, Norvin Green reportedly boasted that it had been "the best financial investment ever made" because managers were able to extract a third more work out of the chastened operators, allowing the company to recover the strike's cost in six months. Perhaps with this remark in mind, Chairman Henry W. Blair badgered Green to give the operators a badly needed raise. Throwing back Green's testimony about Western Union's superiority as a "money-making instrumentality," Blair urged, "Now that you have got your own way about it I wish you would just come up and give those boys more money."[66]

Animus toward Gould and Western Union finally took legislative form in the winter of 1884. In January Colorado senator Nathaniel P. Hill introduced a bill to buy out the existing telegraph companies at a fair valuation under the 1866 act. If Western Union resisted, Hill advocated the construction of a competing postal network. He thought that a postal telegraph could handle public telegrams for a penny a word as an "enlightened public service" instead of a for-profit business conducted by a monopoly with swollen capital. Although Hill's bill never came to a floor vote, the internal correspondence of Norvin Green and other Western Union managers shows that they were more concerned about it than any other piece of postal telegraph legislation since the early 1870s. Green, for instance, wrote letters to several Republican senators calling Hill's bill "a plan of confiscation." He enclosed a list of shareholders in their states to reinforce his oft-repeated assertion that widows, orphans, and others of small means owned Western Union stock. Green also attributed Hill's "rampage" to his "antagonism" to Gould.[67]

One reason for the failure of Hill's bill was behind-the-scenes legislative maneuvering on the part of Norvin Green and Jay Gould. A deeper reason was the renewal of telegraphic competition. At the same time that the Senate Post Office Committee was considering Hill's measure, the Baltimore and Ohio Railroad embarked on an ambitious plan to use its vast rail network to set up a competing telegraph company. Later that summer, the Baltimore and Ohio Telegraph Company joined with the Bankers and Merchants and Postal Telegraph companies to set up a united front against Western Union. This alliance lasted only a few months because the Bankers and Merchants went bankrupt in the fall. The B&O found itself overextended as well. Under the leadership of the impetuous Robert Garrett, the railroad company had embarked on an aggressive plan to operate its own express company and passenger-car factory as well as a telegraph network. At the behest of J. Pierpont Morgan, the B&O spun off these ancillary activities in 1887. The Postal Telegraph Company, however, was able to maintain itself as a competitor to Western Union. In the fall of 1887, the two companies agreed to equalize their rates and Postal remained in the field until World War II.[68]

In 1892 Walter Clark, a justice on the North Carolina Supreme Court, thought that the postal telegraph would remain a leading public issue until Congress passed a measure. "Like Banquo's ghost," he concluded, "it is a question which 'will not down.'" But it had already become clear that the heyday of the movement was past. The last significant postal telegraph plan was Postmaster General John Wanamaker's 1890 proposal to connect the nation's post offices with a network of leased wires on a model similar to the wire leases Western Union had with press associations and brokerage firms. Although Wanamaker's plan resulted in extensive hearings and press coverage, it ultimately led nowhere. The Populist, labor, and public ownership movements continued to press for postalization for the next several years to no avail. In 1896 William L. Wilson, Grover Cleveland's postmaster general, concluded that there was "no great popular demand" for a postal telegraph and that it was "scarcely probable" that Congress would pass such a measure. In 1903 a commentator in the *Outlook* noted that it was no longer "generally known" that the government had the right to purchase the nation's telegraph lines under the 1866 act.[69]

Even an eighty-nine-day telegraphers' strike in 1907 could not make the postal telegraph a leading political issue again. In an editorial written just after the start of the strike, the *Independent*, though it thought that government operation of the wires would eventually result from gradual "industrial and political evolution," concluded that "a serious war would probably bring about the change at once." The journal was half correct. The Wilson administration took control of the wires during World War I, but federal control proved so disastrous that it permanently ended the postal telegraph movement and dampened the movement for public ownership of utilities generally.[70]

During the Wilson administration, advocates for government control of the wires assumed a decidedly Progressive tone. Before the turn of the century, reformers had passionately called for a postal telegraph to eliminate the abuses arising from private monopoly, employing the heated antimonopoly rhetoric characteristic of the day. Instead, Wilson and his cabinet relied on calm statistical analysis and sober claims of efficiency to make their case. Whereas postal telegraph reformers of the nineteenth century said little about the telephone, their Progressive successors were determined to merge all the wires into the postal system. Wilson, at least six cabinet members, and his personal adviser Joseph Tumulty all supported postalization of the telegraphs and telephones. On the national stage, public ownership of utilities was a central focus of Progressive reform, and it seemed likely that the Post Office would eventually assume control of the wires just as many cities had taken over street traction and electric power plants.[71]

A few weeks after taking office, Wilson privately wrote his postmaster general Albert S. Burleson, "For a long time I have thought that the government ought to own the telegraph lines of the country and combine the telegraph with the post office. How have you been thinking in this matter?" Publicly, however, Wilson denied that he was considering a postal telegraph and left the matter to Burleson, one of the strongest supporters of a postal telegraph in a generation. Burleson issued a lengthy, statistics-laden report in January 1914 attempting to prove that a government-operated telegraph and telephone system would be more efficient than the presently privately owned networks. In Congress, Maryland representative David J. Lewis attempted unsuccessfully to secure passage of Burleson's plan in 1914 and 1915.[72]

Federal control of the telegraphs and telephones did not arrive as the gradual realization of Progressive reform but came under the cover of an emergency war measure. Private diaries and correspondence of Wilson's cabinet members reveal that they regarded the war as an opportunity to push through their plans for federal control of "all methods of communication," as Navy Secretary Josephus Daniels phrased it in late 1917. The precipitating event was a threatened telegraphers' strike in July 1918 that would have severely hampered wartime communications. Yet, as Wilson put it to Senator Morris Shepard in late June, his intention to nationalize the telegraphs was "not immediately connected with the conduct of the war." Daniels, Secretary of War Newton Baker, and Burleson also made it clear to Congress that they sought permanent—not just wartime—federal control of the telegraph, telephone, and wireless systems. Congress consented to give government control a wartime trial and passed legislation authorizing Wilson to take over the telegraphs and telephones on 1 August 1918. Wilson promptly formed a Wire Control Board that set rates and wages for the communications industry, though telegraph and telephone companies were never formally integrated into the Post Office.[73]

During testimony on this legislation, Burleson asked Congress to make wartime administration a thorough test of public ownership of the wires. He was optimistic that Post Office administration would prove so successful that the public would demand that it be made permanent. He claimed that postal administration would slash telegraph and telephone operating costs by 25 percent, some $110 million annually, thus reducing customers' rates by that amount. Congressman David J. Lewis also promised that combining the telegraph and telephone systems would double the message-handling capacity of the telegraph network at a stroke.[74]

These overly optimistic predictions were far off the mark. Burleson had predicted that government operation would satisfy all parties—customers, stockholders, managers, and employees. He promised to lower rates while raising wages, for example, never acknowledging that these goals were contradictory. Instead, Burleson was forced to raise telegraph and telephone rates some 20 percent to cover these wage increases. Unions remained hostile as well, rightly regarding Burleson as an enemy of organized labor. As criticism mounted, Burleson became increasingly autocratic, alienating many of his erstwhile supporters. For example, in November 1918, several weeks after the Armistice, Burleson summarily replaced the upper management of the Postal Telegraph Company. In the spring of 1919 Burleson barred Postal's managers from administering overseas cables as well, because of their refusal to increase wages and shorten working hours. Although President Clarence Mackay and General Manager Edward Reynolds had been remarkably uncooperative, many observers regarded Burleson's acts as despotic.[75]

Whatever Burleson's faults as an administrator, he seems to have had the gift of prophecy. When Burleson sought to convince Congress to authorize postal administration of the wires in the summer of 1918, he predicted that this would constitute a practical test of the desirability of public ownership of utilities. If wartime operation failed, he claimed, "that would end it." By April 1919, the newspaper press of the country—of all political stripes—unanimously concluded that Burleson had singlehandedly killed the public ownership movement. In response to mounting criticism, Burleson reversed his statements of the summer of 1918 and maintained instead that wartime operation did not constitute a fair test of government ownership. He stubbornly continued to press for government control, although it was clear that the rest of the Wilson administration, the Republican Congress, and the public generally had grown weary of the experiment. Tumulty advised Wilson in May 1919 to return the wires promptly to private control, because government administration had "produced nothing but criticism and discontent. The sooner we can get rid of them, the better." At the end of July, the government returned the wires to their corporate owners. Never again would the public seriously consider government control of communications, even during wartime. Just as the Wilson administration's wartime overreach helped to end the Progressive movement writ large, disastrous government control of the wires was a major setback for the public ownership movement.[76]

Although reformers never succeeded in placing the telegraph under permanent government control, their attempt to do so was an important part of Gilded

Age and Progressive Era political culture. Between 1866 and the turn of the century, every Congress except the 44th (1875–77) debated measures to postalize or regulate the telegraph. Committees in both houses reported bills nineteen times, seventeen times in favor and only twice against. Postal telegraphy garnered widespread support from labor unions, farmers' organizations, the Populist Party, the National Board of Trade, several dozen local chambers of commerce, much of the newspaper press, and prominent politicians, including presidents Ulysses S. Grant and Chester A. Arthur and most postmasters general. In 1888 Congress received a petition to nationalize the telegraph signed by more than 2 million people. Economist Frank Parsons, a leading advocate of nationalization in the 1890s, estimated that business, agricultural, and labor organizations with a total membership of 24 million people supported the postal telegraph movement at the turn of the century.[77]

Yet the movement's leaders were unable to transform this impressive numerical strength into legislation. Reformers presented a bewildering array of alternatives because they could not agree on the proper limits of federal authority. Contrast the nationalization of the British telegraphs with American efforts. The British nationalized the telegraph in the late 1860s with little or no debate over the limits of parliamentary authority to do so. The American federal government, on the other hand, had clearly defined limits on its authority to regulate both interstate commerce and state-chartered corporations. In addition, the Post Office Department was a major source of patronage for the party in power, making an expansion of its duties unpalatable to many legislators. Thus, many Americans—even telegraph reformers—questioned the wisdom and legality of a government-owned telegraph system. These concerns explain why telegraph reformers were unable to coalesce behind a particular plan.

At the same time, the telegraph's major customers, the press and business, were mainly satisfied with private ownership. They turned against Western Union's monopoly for a few years in the early 1880s, after Jay Gould's 1881 takeover and the 1883 operators' strike, but business and press support for a postal telegraph receded after competition reemerged in 1884. Thus, many members of Congress saw no reason to interfere with an industry that apparently served its major customers well.

More generally, the postal telegraph movement ebbed and flowed with the presence and absence of meaningful competition with Western Union. The movement peaked at periods when private competition seemed hopeless. The movement came closest to succeeding around 1870 and 1883, periods when Western Union enjoyed almost total dominance of the industry. But when competitors arose to assume the antimonopoly mantle, postal telegraphy quickly faded from public view.

A final reason for the movement's failure was that support for a postal telegraph was broad but not deep. While organizations such as the Knights of Labor, American Federation of Labor, and the Populist Party backed a postal telegraph, they were focused on issues of more immediate and vital concern to their members. Support for the postal telegraph was also halfhearted because only about one in sixty Americans had occasion to use the telegraph. There is irony in this, because this low level of usage was a central reason why reformers urged nationalization and also why the movement failed to gain traction.

Because the postal telegraph movement failed, it has all but disappeared from historical notice, except among academic specialists. This is unfortunate because the movement was a major episode in American politics. It deserves to be remembered for that reason alone. Between 1866 and 1900, Congress considered nearly one hundred bills and issued forty-eight committee reports on postal telegraphy, making it one of the most heavily studied and debated issues of the day.[78]

The postal telegraph movement was important for deeper reasons as well. Although the private operation of telephones and telegraphs might at first glance support the notion of American exceptionalism, a deeper examination in fact reveals the transatlantic character of contemporary reform movements. Just like the cheap postage movement of the preceding generation, the postal telegraph movement drew inspiration and support from the British nationalization of the telegraphs in the late 1860s.

Similarly, the postal telegraph movement calls into question the traditional division of the era into a Gilded Age, lasting from 1865 to about 1890, when Americans first became aware of social and economic problems and a Progressive Era, from about 1890 to World War 1, in which Americans aggressively moved to solve these problems.[79] Instead, the postal telegraph movement showed remarkable continuity over its fifty-year life that spanned this common periodization.

The postal telegraph movement was central to the rise and fall of the larger public ownership movement. The concept of natural monopoly provided this linkage. Although the postal telegraph movement drew its initial inspiration from the successful cheap postage movement of the previous generation, reformers also argued that a networked technology like telegraphy tended inexorably toward consolidation. After the late 1880s, economists took this assertion further to give full articulation to the notion of natural monopoly. By the turn of the century public ownership advocates claimed that natural monopolies were in fact "public utilities" that ought to be controlled or at least closely regulated by the state. The public ownership movement did not outlive the disastrous postal operation of the wires during World War I.[80]

The postal telegraph movement left a far deeper imprint on American political economy than simply as a failed episode in reform. In retrospect, the continued private ownership of telecommunications networks was instrumental in establishing the boundary between public and private enterprise and in erecting the twentieth-century regulatory state,. More importantly, federal abandonment of electrical communication to the private sector might well be the origin of today's cheapened political debate and impoverishment of the public sphere. These concerns prompt us to consider again Samuel Morse's words from 1837, whether private ownership of the means of communication has been "an engine for good or evil."[81]

CHAPTER THREE

"There Is a Public Voracity for Telegraphic News"

THE TELEGRAPH, WRITTEN LANGUAGE, AND JOURNALISM

In 1848 a contributor to the *United States Magazine and Democratic Review* speculated on the telegraph's "probable effect upon Style in Composition," concluding that the "certain effect of the Telegraph . . . will be to introduce a style of writing which shall be *first of all, brief,* . . . terse, condensed, expressive, sparing of expletives and utterly ignorant of synonyms." The newspaper press was to be the mechanism for inaugurating this "Yankee directness" into prose composition generally: "When a half column or more of every paper in the Union is filled with Telegraphic dispatches; when these reports form a large part of the daily reading of thousands; when correspondence is hourly prepared and revised, throughout the whole extent of the United States, with a view to telegraphic transmission, is it too much to expect that this invention will have an influence upon American literature; and that that influence will be marked and permanent, and withal salutary?"[1]

Ernest Hemingway, embarking on his literary career some seventy-five years later, seems to prove this prediction. As a European cable correspondent for the *Toronto Star* in 1922, Hemingway immediately took to "cablese," the stripped-down language used to file stories for overseas transmission. His colleagues George Seldes and Lincoln Steffens recalled that Hemingway "came in one night and said: 'Stef, look at this cable: no fat, no adjectives, no adverbs—nothing but blood and bones and muscle. It's great. It's a new language.'" After that, Seldes claimed, "Hemingway's writings . . . are considerably changed."[2]

The anonymous writer in 1848 and Hemingway's colleagues in the early 1920s concluded that the telegraph had the power to transform written language, particularly literary prose. At first glance, this is unsurprising because the telegraph was the first technology to interact with the written word since the invention of the printing press in the fifteenth century. Both accounts also suggest that the most fruitful realm for investigating this relationship is journalism, the site at which the telegraph and writing interacted most strongly.[3]

However, the effects of telegraphy on journalism and written language are more complex than these accounts indicate. This relationship can be approached from three directions. To begin with, the technology and economics of telegraphy directly enforced brevity in composing and transmitting telegrams. This brevity imposed by the medium's cost had a further effect—it changed the production and consumption of news. The press was one of the earliest, most enthusiastic, and largest customers of the telegraph. After the Civil War, the press accounted for 15 to 20 percent of Western Union's message traffic. Tellingly, both the New York Associated Press and Western Associated Press located their headquarters within Western Union's headquarters.[4] Just as importantly, telegraphic newsgathering changed the way Americans consumed the news, helping to create modern sensibilities about timeliness and newsworthiness. Though telegraphic brevity engendered profound changes in the news, it is unlikely that it changed prose composition generally. Distinguishing between the telegraph's social roles as both a communication and an information medium explains why. Because only a small minority of Americans used the telegraph to send or receive messages, telegraphy's effect upon written language was indirect and ambiguous at best. Most Americans encountered the telegraph as a source of information in the pages of their daily newspapers. However, they did not usually read telegraphic dispatches directly—newspaper editors typically used these dispatches as raw material to write longer and more detailed stories.

Telegraphy's Direct Linguistic Effects

From the beginning of the industry, the economics of telegraphy acted to condense written language. The key point here is the relationship between a telegraph company's message-carrying capacity, its marginal cost to send an additional message, and the price it charged customers. Unlike methods for transporting mail, a telegraph line could transmit only one message at a time. This resulted in a different marginal cost structure from the postal system's, one based on time or message length rather than on weight or size. For this reason alone, managers sought to reduce the amount of time required to send messages. Also, from an operational standpoint, competition among operators for use of a telegraph line was a common problem in the early industry. The faster a station could clear its outbound traffic, the sooner other stations could have access to the line. Condensed messages also improved capacity utilization, allowing more messages to be sent for the same fixed and working capital costs. Finally, telegraph companies charged their customers per word—as written out on the message blank, not as actually transmitted. If the sending operator eliminated or abbreviated words so as to condense a hundred-character dispatch into fifty letters, the company correspondingly reduced its cost by that proportion. Thus, the economics of telegraphy meant that shorter messages reduced costs and increased profits.

This condensation of messages was integral to the telegraph's operation from its beginning. Samuel Morse patterned his eponymous code after the statistical distribution of English letters, assigning the shortest codes to the most frequently occurring letters. The most common letter in English text, *E*, received the shortest code, one dot. All of his letters, regardless of relative frequency, were assigned codes that could be transmitted within the time it took to make five dots. Morse explained that his code conformed "to the capacity of my instrument," its basic "principle" being the "consumption of as little time as possible in writing."[5]

From the beginning of commercial telegraphy, operators dropped words of low information content and abbreviated long words. A few days after the first Washington to Baltimore line opened for business in May 1844, Morse (at Washington) directed his colleague Alfred Vail (at Baltimore) to "condense your language more; leave out 'the' when ever you can," and to send only the "beginning of a long common word." In the fall of 1844 Morse demonstrated his telegraph to a group of people, including two or three congressmen who had voted for his appropriation and a young Cincinnatian, Cyrus Mendenhall. Mendenhall recalled that Morse sent "WHAT NEWS HAVE YOU?" to Baltimore and received this verbatim reply: "IT IS FEARED THERE WILL B A FLOOD IN JONES FALLS AS I UN THRO T CITY." Morse's decoding showed that B meant "be," UN "understand," THRO "through," and T "the." A few years later, another telegrapher wrote Vail that "all articles, prepositions of and under four letters, and the verb to be," are either "omitted" or replaced "by a close condenser. Pronouns of four letters and under . . . are also frequently omitted." The goal of such abbreviation was to convey information in as few characters as possible. In 1845 Vail told Theodore Faxton, president of the newly formed New York, Albany, and Buffalo Telegraph Company, that operators eliminated nearly one-half of the characters customers wrote down on message blanks.[6]

After sound reception became common in the early 1850s, abbreviations and codes became staples of operators' work culture, a type of specialized language they delighted in to set themselves apart as practitioners of an arcane craft. One of the most common and expressive codes, "i i," (the letter *i* being two dots) was a snappy shorthand for "Aye, Aye," "I am ready," or "I will carry out your instructions."

The number *1* was shorthand for "Wait one," or "Stand by for a moment." The closing *73* stood for "Best regards." Operators not only used these codes over the wires but also included them in official company correspondence, in personal letters to each other, and in articles in telegraphers' magazines.[7]

After the industry stabilized in the late 1850s, companies turned their attention to improving reliability and speed of message delivery and to increasing their message-carrying capacity. Both considerations affected the industry's cost structure. Marshall Lefferts, engineer of the American Telegraph Company, astutely noted the relationship between transmission speed, price, and message length. During the Civil War, he reported to the company's directors that customers expected "a more prompt and reliable transmission of despatches. . . . A few years since . . . messages delivered within a few hours of their reception would have caused no complaint. Now we are held to minutes instead of hours." To satisfy demanding customers like businessmen and the press, he recommended three classes of messages, a priority service at a premium of fifty cents over standard rates, using dedicated wires and the best operators; a regular service at existing rates; and a reduced-rate service for messages of twenty to thirty words transmitted after regular business hours for delivery the next morning. He thought that the latter service would encourage "a new class of business—a feature in commercial life of writing by telegraph instead of the mail, and at a time of the day when our regular business begins to slacken."[8]

The American Telegraph Company did not implement Lefferts's recommendation for a priority service, nor did Western Union after it bought out the American in 1866. Although Western Union experimented with a lower-rate, deferred night-service between 1867 and 1882, its managers consistently rejected a higher-rate premium class of business. Instead, it continually and gradually improved service for all its customers by reducing transmission and delivery times and gradually reducing rates. Not only did these improvements better serve its main classes of customers (businessmen, financiers, and the press), but they also helped forestall competition and blunt the appeal of a government telegraph system. Two technological innovations in the 1870s were essential in increasing message-carrying capacity and reducing rates. In 1875 the company acquired the rights to the quadruplex, a device invented by Thomas Edison for simultaneously transmitting four messages on one circuit. At the same time, continued improvements in automatic repeaters meant that messages traveled longer distances without retransmission by hand and that traffic managers could make up and cut over long-distance circuits more quickly and reliably. The vast increases in bandwidth permitted by these technologies allowed Western Union to reduce its rates by half from 1868 to 1888 while doubling the number of messages handled and maintaining the same profit margins.[9]

Despite Western Union's embrace of these technological innovations, its marketing remained remarkably hidebound until the 1910s. Its managers did not attempt to popularize telegraphing and resisted any modifications to its cost structure, based on a standard tariff of ten words plus an additional charge for each additional word. The company's management continued to insist that the telegraph was an expensive medium used mainly by businessmen and the press, customers who willingly paid well for rapid transmission of dispatches. Ordinary people rarely used the telegraph and continued to rely on the postal system for the bulk of their long-distance communications. Indeed, most Americans regarded telegrams as harbingers of death or misfortune. For example, in Louisa May Alcott's celebrated Civil War novel *Little Women*, the March family receives word of Mr. March's grave illness through "one of them horrid telegraph things." The family's maid "handl[ed] it as if she was afraid it would explode," and after Mrs. March read its contents, she turned "as white as if the little paper had sent a bullet to her heart."[10] The First and Second World Wars strengthened this connection between telegrams and disaster, because families received official notice of combat casualties this way. This remained a major public relations and marketing problem for Western Union after 1920.

Statistics from the nineteenth century amply demonstrate that the vast majority of Americans rarely sent or received telegrams. More than 70 percent of the message traffic in the early 1850s was commercial in nature, while less than 10 percent consisted of social or family messages. In 1851 the superintendent of the New York and New England Telegraph Company put the proportion of commercial message traffic at two-thirds to three-fourths. George Thurston, the president of the Pacific and Atlantic Telegraph Company, told a congressional committee in 1873 that only one in four hundred residents of Pittsburgh had occasion to send even one telegram a year. Norvin Green, president of Western Union, told Postmaster General William Vilas in 1887 that less than 2 percent of the American population ever used the telegraph and that 87 percent of Western Union's revenue came from business messages (the majority from stock and commodity speculators and racetrack gamblers). The press accounted for 8 percent of the company's revenue (mainly because of sharply reduced rates), and only 5 percent of revenue came from social and family messages.[11]

Only after the American Telephone and Telegraph Company (AT&T) acquired control of Western Union in 1909 did its management embrace Lefferts's fifty-year-old idea of popularizing telegraphy by setting up tiered classes of service.

The company had started a night-message service at two-thirds full-day rates in 1867, but for reasons that are unclear, it deliberately discouraged this class of business after 1882 by keeping night rates steady while continually reducing day rates. Only in 1910, with a more progressive management installed by AT&T, did Western Union inaugurate a Night Letter service, charging the same rate for a fifty-word telegram sent at night as it did for a regular ten-word telegram sent during business hours. By 1912, Western Union handled some 15 million Night Letters, marketing them to both business and social customers. In the company's annual report for that year, president Theodore Vail adopted the new language of public-utility economics to claim that these new off-peak services met "the varied demands of business and social communication," gave "the greatest possible benefit to the public," provided a "fair return," and "modif[ied] the abrupt traffic curves" that had characterized the telegraph business. Or, more succinctly phrased: "It is an axiom that full loads make cheap rates."[12]

Although AT&T divested itself of Western Union under pressure from federal antitrust regulators in 1914, AT&T was able to modernize its plant and marketing philosophy during its brief period of control. Western Union began switching over from Morse operators to high-speed multiplex printers that vastly increased its message-carrying capacity.[13] By the onset of the Great Depression, Western Union's modernization program was largely complete, just in time for business message traffic to drop sharply. Excess bandwidth and slumping revenues led Western Union's managers to embrace sociability fully and to market its services more aggressively. During the Depression the company rolled out so-called Fixed Text messages, canned birthday and holiday greetings. One purpose of these messages was to counteract the prevailing negative image of the telegram as a harbinger of disaster and to encourage the habit of telegraphing for happy occasions. Western Union marketed these twenty-five-cent greetings with the slogan, "Isn't there someone who would like to hear from you today?" Customers simply checked a box on a message blank and the company transmitted a number standing for the particular message chosen. For example, "FT400" meant, "I'm just a little tot, I haven't much to say, just want to wish you a happy Mother's Day." By 1940 Americans sent some ten million of these messages a year.[14] From a linguistic perspective, these canned messages were the logical culmination and fullest extent of compression driven by telegraph industry technology and economics.

From the beginning of the telegraph industry, therefore, company managers sought to compress messages in order to reduce costs. Customers, charged by the word, likewise condensed their dispatches as much as possible. Modifiers, prepositions, and articles evaporated, leaving messages consisting of nouns, verbs, and

TABLE 3.1.
Message Length per Telegram Sent from New York, October 1903

Paid Words per Telegram[a]	Messages (%)	Cumulative Percentage
2–9	1,064 (35.3)	35.3
10	630 (20.9)	56.2
11–15	797 (26.5)	82.7
16–20	286 (9.5)	92.2
21–25	113 (3.8)	96.0
>25	121 (4.0)	100

SOURCE: Statistical Notebooks, WUTC.
 Note: Sample includes 3,011 telegrams. More than half of the messages were 10 words or less; average message length was 11.93 words.
 [a] No 1-word telegrams were sent.

numbers. Users eliminated the typical rhetorical conventions found in written correspondence. Repetition, for emphasis or for clarity, became redundant and wasteful. Niceties of style, elegant turns of phrase, and formal greetings and closings were telegraphic extravagances. In 1860 telegraph electrician and manager George Prescott advised customers to write telegrams "in a concise style," with "no superfluous words employed." Because telegraph companies charged a fixed rate for the first ten words, then additionally for each word thereafter, most telegrams contained ten or fewer words. Prescott noted that "it is surprising how much matter is often contained in this brief number." Similarly, a 1914 business English textbook challenged students to draft ten-word telegrams from business letter passages of twenty to thirty words. An exercise in a 1915 textbook directed students to "Prepare a want-ad or telegram. Avoid complete sentences. Aim at the utmost compression." The humorist George Ade had this compression in mind when he quipped in 1905, "A man never feels more important than when he receives a telegram containing more than ten words."[15]

Telegrams retained in business and government archives reveal this relationship between user's cost and the length and importance of telegrams. For example, a sample of about fifty telegrams sent to the Cincinnati merchant M. Bare & Company in 1866 shows results similar to the Western Union statistics collected in 1903. The average length of telegrams received by the firm was nine and two-thirds words, with about two-thirds of the telegrams having a word count of ten or less. Only half (eleven of twenty-two) of the telegrams sent collect (paid by recipient) were ten words or less with an average word count of just over eleven words; in contrast, twenty-two of twenty-seven telegrams paid by the sender were ten words or less with an average word count of just above eight.[16]

Military telegraphers during the Civil War noted a similar connection between user's cost, a telegram's length, and the message's importance. Commanders at the brigade level and higher typically had a military telegrapher attached to their staffs, and were able to send telegrams as often as they wished at no personal cost. Union telegraphers frequently criticized officers for overuse of the telegraph. For example, Captain William Gross of the U.S. Military Telegraph Corps (USMT) complained in October 1864 that he was unable to send important military messages promptly because some generals "conduct[ed] their entire correspondence by telegraph" when messengers or the mail would suffice. They clogged it with "long unimportant dispatches at a time when the utmost celerity is demanded for really important dispatches." To Gross this was "a perversion of its use. . . . The prevalent idea that he who sends the most dispatches is more efficient, is as untrue as it is absurd."[17]

The experience of the British telegraph system after its 1870 nationalization showed a similar relationship between user cost, length, and importance. Seeking to emulate the success of the penny-post movement of the 1830s and 1840s, the British Post Office made sharp reductions in telegraph tolls to transform the telegraph from a business medium to a social medium. Immediately upon taking control of the telegraphs, the Post Office slashed rates for a twenty-word message from three to four shillings to one shilling. As critics often pointed out, the result was a flood of frivolous and wordy telegrams. Economist William Stanley Jevons, who had been an advocate of nationalization before 1870, reconsidered his position five years into the experiment. He thought that much of the increased traffic resulting from the rate reductions consisted of "complimentary messages, or other trifling matters. . . . Men have been known to telegraph for a clean pocket-handkerchief." Similarly, in 1898 Henry James in his novella *In the Cage*, criticized wealthy Britons' "profligate" use of the Post Office telegraph, "the 'much love's, the 'awful' regrets, the compliments and wonderments, and vain, vague gestures" that flitted over the wires. And the prolix yet inconsequential telegrams exchanged between Bertie Wooster and his bothersome aunts were running jokes throughout P. G. Wodehouse's *Jeeves* stories of the 1920s and 1930s.[18]

American telegraph customers, however, had become so habituated to brevity in composing telegrams that they found it hard to be prolix even when cost constraints were removed. In 1915, for instance, AT&T considered competing with Western Union for Day and Night Letters, options that allowed customers to send thirty- or fifty-word telegrams for deferred delivery at the same rates as ten-word telegrams transmitted immediately. An internal AT&T memorandum noted that the average number of words in the fifty-word Night Letter was only forty, concluding that "the public is not availing itself of all the words it is entitled to send for the rate charged." Americans had come to regard telegrams as media for brief and urgent communications. Any message of fifty words could have probably been mailed just as readily and more cheaply.[19]

Ocean cables were much more expensive to install and operate than landlines. They also had far less message-carrying capacity because of signal distortion and attenuation. Until the development of inductively loaded cables in the 1920s,[20] transmission speeds rarely exceeded twenty words per minute. Thus the economics of cable telegraphy enforced much stricter brevity. After failed attempts in 1857, 1858, and 1865, the Anglo-American Telegraph Company succeeded in laying two permanently working cables in the summer of 1866. High rates over the cables, initially $100 for twenty words, encouraged maximum linguistic compression. These high rates led to early and frequent use of ciphers among financiers and the press, the two leading customers of the cables. In 1870 financial writer James K. Medberry estimated that New York stock and commodity dealers paid about $1 million annually for London dispatches at that rate, even in cipher. To help financiers reduce their transmission costs, a former manager of the American Telegraph Company, George Stoker, began a lucrative business as a "cable packer." Stoker used a cipher that "packed" several messages into one twenty-word message. By 1877, Gold and Stock Telegraph Company, Western Union's subsidiary handling its ticker business and financial news reporting, was earning $10,000 a year from cable packing. In 1902 George G. Ward, vice-president of the Commercial Cable Company, estimated that 95 percent of cable messages were enciphered, one transmitted word expanding to five to twelve words when deciphered.[21]

As with landline telegraphs, the press was an early and heavy user of the Atlantic cables. In 1857, when work had begun in earnest on the first Atlantic cable, the *New York Times* expected to carry a thousand words of European intelligence per day, at the expected rate of $1 per word. A year later, when a cable worked briefly, the *New York Ledger* planned to hire Charles Dickens to transmit an original story, at great expense, over the cable. The *New York Herald*, on the other hand, thought that the cable would result in little effect on "literary pursuits," since "the quality of fastness is rather opposed to the pleasure of deliberate and critical enjoyment." After the 1866 cable expedition succeeded in laying two permanently working cables, the *New York Herald* paid $7,000 (about $90,000 in current value) to obtain an important speech by King Wilhelm of Prussia.[22]

Given these high costs of cable transmission, correspondents tried to compress messages as much as possible. George Smalley, the *New York Tribune*'s London correspondent, recalled that cable news had a "peremptory brevity which arrested attention. The home telegraph was diffuse. It was the cable which first taught us to condense." In early August 1866, Smalley sent what he thought was the first press dispatch sent over the Atlantic cable, a hundred-word dispatch costing $500 detailing Prussian troop movements against Austria. "We wasted no words at that price," Smalley remarked. A New York Associated Press dispatch during the Franco-Prussian War four years later serves as a good example of this linguistic compression. The New York office received this puzzling dispatch after a major battle: "Salmsafn Major 46 regiment Prussian Guards killed 186." At first glance, the Associated Press agent in charge of working cable dispatches into brief news stories thought that a Major in the 46th regiment of the Prussian Guards killed 186 French troops singlehandedly. After some detective work, the agent discovered that there was no 46th regiment, concluding that "Salmsafn" was in fact "Salm Salm" and the number "46" was instead "4th." The resulting newspaper story, based on this garbled eight-word dispatch, read: "Prince Felix Salm-Salm, distinguished for his services in the American and Mexican Wars, was killed in the battle of the 18th at Gravelotte. He was a major in the 4th regiment of Prussian Grenadiers of the Guards, and fell at the head of his men." Later stories fleshed out this bare-bones item with details of Salm-Salm's marriage to an American woman and his service in the Union army during the Civil War.

Some fifty years later, William Shirer gave an example of a "cablese" dispatch detailing a dispute between France and Germany over First World War reparations payments: "EXCLUSIVE POINCARE CHICATRIBWARD UNTRUTH UNPAY WARDEBTS AMERICAWARD STOP FRANCE UNINTENDS UPGIVE REPARATIONS DUE EXTREATIES STOP TWO LINKED QUOTE UNPREPARATIONS UNWARDEBTS PAYABLE UNQUOTE UNBELIEVES GERMANS UNFUNDS PAY FULLEST STOP UNBEFORE POINCARE ADAMENTEST REGERMANS DELIBERATE STALLING STOP BRISTLING CHICATRIBWARD UNEXCUSES EXGERMANS." This 39-word dispatch not only gave enough details to become a 144-word story but also contained routing information that the story was an exclusive for the *Chicago Tribune*. These examples show that highly condensed cable dispatches, even those sent in clear text, required judgment, creativity, and some embellishment to expand them into news stories.[23]

The Telegraph and the Newspaper Press

The telegraph influenced American journalism in three major ways.[24] The high cost of collecting and distributing news by telegraph led to its centralization in the hands of press associations like the New York Associated Press (NYAP) and later its sister regional press associations like the Western Associated Press

(WAP). Although press associations and telegraph companies had a contentious relationship before the Civil War, in the last third of the century many Americans feared that the nation's flow of news had fallen into the hands of a joint Associated Press–Western Union monopoly. Because press associations sold their news to newspapers of all political stripes and editorial philosophies, newsbrokers instructed their correspondents to provide only facts and to transmit the most important facts first. Thus, the matter-of-fact style and inverted-pyramid structure of the modern news story originated with telegraphic newsgathering and reached a level of maturity during the Civil War. In addition to changing the business of newsgathering, the telegraph changed the psychology of news consumption. By permitting frequent updates to evolving news stories, from the Civil War onward the telegraph helped create modern expectations of timeliness and newsworthiness. During the late nineteenth century, news-hungry crowds routinely gathered at hotels, newspaper offices, and telegraph offices to keep up with the latest updates on major news stories.

A close, if not always harmonious, connection between the telegraph and the press existed from the very beginning of the American telegraph industry. Even before Samuel Morse officially opened his Washington-to-Baltimore line, he and Alfred Vail had already begun reporting the proceedings of the Whig Party convention in Baltimore. Morse's inaugural message that officially opened the line on 24 May 1844 was the Old Testament passage "What hath God wrought!" His lesser-known second message was, "Have you any news?" Over the next few months, Morse commonly asked Vail for Baltimore news as a way to demonstrate his telegraph's capabilities to curious visitors. On 25 May Morse began transmitting summaries of congressional proceedings to the *Baltimore Patriot*. A few days later, Alfred Vail sent Morse frequent updates on the progress of the Democratic convention in Baltimore, further demonstrating the telegraph's utility as a newsgathering tool.

A few newspapers, especially James Gordon Bennett's *New York Herald*, were early and enthusiastic adopters of the telegraph. However, many other newspapers were reluctant to pay for telegraphic news. Daniel Craig, one of the founders of the NYAP, recalled that, when he tried to organize newsgathering on a national scale, "a majority of the editors [were] violently opposed" to his plans. Organized wire-service journalism owed its origins to the fierce and expensive competition between New York newspapers for European steamer news and Mexican War news in the late 1840s. At that time the Morse lines charged $0.50 for a 10-word dispatch from Boston and $2.40 for 10 words from New Orleans, with no rebates given to the press. The entry of new lines using the Bain and

House telegraph systems led to competition that reduced these rates substantially in the 1850s. Still, newspapers continued to rack up large telegraph bills. To economize on their costs, a consortium of New York newspapers formed the New York Associated Press in 1848. To get a sense of the expenses incurred, the NYAP paid $500 in telegraph tolls per arriving steamer to transmit European news from ships passing Halifax, Nova Scotia, to New York. To transmit the 1848 election returns from around the country, the NYAP paid telegraph lines more than $1,000. The NYAP's first general agent, Alexander Jones, estimated in 1852 that the organization paid telegraph lines $25,000 to $30,000 a year and that press reports accounted for 3,000 of the roughly 7,000 words transmitted per day by the New York, Albany, and Buffalo Telegraph Company. Within twenty years, these figures increased exponentially; in 1869 Western Union handled some 370 million words of press matter at a cost of about $884,000 (or just under a quarter cent per word).[25]

To save on NYAP's telegraph bills, Jones developed a cipher to facilitate transmission of commercial news from six cities to New York. This cipher used words of three to five letters to stand for much longer stock phrases. Jones gave an example of a nine-word cipher telegram that, when deciphered, expanded to a market report from Buffalo of sixty-eight words: "Flour market for common and fair brands of western is lower, with moderate demand for home trade and export. Sales 8,000 bbls. Genesee at $5.12. Wheat, prime in fair demand, market firm, common description dull, with a downward tendency, sales 4,000 bushels at $1.10. Corn, foreign news unsettled the market; no sales of importance made. The only sale made was 2,500 bushels at 67c." Jones developed a similar cipher to transmit legislative proceedings from Albany and Washington, but he discontinued its use in 1850 as telegraph rates dropped owing to competition between lines using rival telegraphs developed by Morse, Royal E. House, and Alexander Bain. Serious mistakes in deciphering these dispatches also played a role in dropping its use. Jones related that his legislative cipher used the word *dead* to mean "[Name], after some days' absence from indisposition, reappeared in his seat." His Senate reporter sent the enciphered dispatch "John Davis dead," upon which several newspapers wrote glowing obituaries of the Massachusetts senator. Davis, of course, was alive and well. In the aftermath, several newspapers commented acidly on the purported "accuracy" of telegraphed news items, leading the president of the Magnetic Telegraph Company to defend his line's operators from charges that they mishandled press dispatches.[26]

NYAP's use of ciphers led to a protracted fight with Francis O. J. Smith, proprietor of the Morse line between New York, Boston, and Portland. Smith, an irascible and disputatious man, decided that the NYAP cipher deprived his line of revenue, so he decreed that he would count every five letters as a word in press dispatches. Jones thereupon drew up a new cipher using five-letter words. Smith then decreed that three letters constituted a word, leading the NYAP to stop transmitting news from Boston and points northward over Smith's line. NYAP telegraphed European steamer news from Halifax to Portland, then sent it by rail to Boston, where it went over the lines of the House telegraph company to New York. For his part, Smith accused the NYAP of attempting to monopolize his line with steamer news to the detriment of other paying customers.[27]

Until the Civil War, relations between the telegraph industry and the press remained fraught. Daniel Craig recalled that the managers of Morse lines were opposed to the formation of a national press association, preferring instead that each newspaper should send its own news dispatches, thus generating more telegraph traffic. The press attempted to assert itself as the industry's most important customer, seeking preferential rates and message transmission priority, while telegraph lines replied that they had to treat all customers alike. At one point in 1859 and 1860, the American Telegraph Company, the major line on the Eastern Seaboard between Halifax and New Orleans, threatened to start up a news service to compete with NYAP. The press association countered with its own demands to choose five of the American Telegraph Company's directors and to have exclusive use of two lines from Washington to New York.

As these demands indicate, NYAP held the upper hand in its dealings with telegraph companies until after the Civil War. In the spring of 1866 Western Union acquired its rivals to become the dominant force in the telegraph industry, and a few months later the Anglo-American Telegraph Co. managed to lay a permanently working Atlantic cable. These two events changed the balance of power between NYAP and the telegraph industry, placing them on an equal footing. If NYAP controlled the collection and distribution of news, Western Union now controlled its transmission.

The Atlantic cable sparked a conflict over the apportionment of the costs of cable news between NYAP and the Western Associated Press (WAP), the most powerful regional association outside of New York City. In January 1867 Western Union forced a settlement between the two press associations. An ominous provision of this three-way agreement was Western Union's insistence that both press associations avoid giving aid and comfort to any competing telegraph line. Supporters of a postal telegraph over the next thirty years correctly charged that this provision dictated the editorial policy of member newspapers, allowing Western Union to shape news coverage of this politically charged question to its advantage.

After Jay Gould acquired control of Western Union in 1881, he briefly considered entering the newsbrokering business. His plan was to set up a new organization, the American News Association, as an equal partnership between Western Union, NYAP, and WAP. He recruited William Henry Smith, head of the WAP, to run this new press association. NYAP refused to enter into this arrangement and instead offered a slightly revised version of the 1867 settlement. However, Western Union's market power proved decisive. In late 1882 Gould forced NYAP's general manager John C. Hueston to step down. Smith promptly took his place at Western Union headquarters as manager of both WAP and NYAP. This arrangement lasted until 1892, when a rival press association, the United Press, started up. NYAP saw this as a final opportunity to regain its former prominence and allied itself with the new association. But the merged association went bankrupt in 1897, leaving the WAP alone in the field. In 1900 WAP rechartered itself in New York as the Associated Press, the modern press association we have today.[28]

A significant provision of the 1882 settlement between Western Union and the two press associations was the increasing use of leased wires. Until this time Western Union had been reluctant to lease circuits, despite frequent requests from both the press and stockbrokers. But the installation of the quadruplex on more and more circuits after 1880 gave the telegraph company excess circuit capacity that it could lease to its customers. The press was Western Union's largest customer, and routing more of its business onto independent circuits meant that the company could provide faster service to its other customers and could devote more bandwidth to its growing stock and commodity ticker business. Western Union's decision to allow NYAP and WAP to lease independent circuits made the company into a provider of circuit capacity as well as a handler of individual messages. Wire services began to operate independent telegraph networks using their own operators and traffic schedules. By the turn of the century, relations between the press associations and Western Union had become arm's length. These wire leases meant that press associations gained a measure of independence from the telegraph company and eased fears, prevalent in the 1870s and 1880s, that Western Union controlled the flow of the nation's news.[29]

The Telegraph and the Changing Structure of News Stories

Wire-service reporting transformed the content and tone of news stories, creating the inverted-pyramid structure consisting of the presentation of the most important facts first. Civil War reporting was an important inflection point in this shift of the structure of news stories from chronological narratives to the inverted pyramid. Congressional hearings on wartime telegraphic censorship and surviving press dispatches in the War Department records provide a unique opportunity to study the technological, organizational, and political factors that lay behind this shift.

Although this transformation would not be fully complete until the 1870s or 1880s, the telegraph began to influence the tone and structure of news stories from the very beginning of the industry. During the first week of the telegraph's operation in 1844, Samuel Morse directed Alfred Vail, transmitting the proceedings of the Democratic and Whig conventions in Baltimore, to "select the most important facts if you are crowded with matter, and leave the rest to be transmitted at leisure." Within five years of Morse's instructions to Vail, telegraph lines covered the Eastern Seaboard from Nova Scotia to Washington, D.C., and extended west to Chicago and St. Louis. The geographic reach of the telegraph network, coupled with the desire to obtain European steamer news and Mexican War news, led several entrepreneurs to establish newsbroking services in the late 1840s. In 1849, for example, Daniel Craig famously stated that he planned to market steamer news from Halifax as a "personal and private enterprise, . . . sell[ing] my news as I would a string of onions, to the highest bidder." The New York Associated Press, formally organized in 1848, quickly edged out its rivals for European steamer news. In 1854, a few years after he had replaced Alexander Jones as general agent of NYAP, Craig issued a circular to correspondents outlining his requirements for telegraphic dispatches. Craig instructed them to send "only the *material facts* in regard to any matter or event, and those facts in the fewest words possible compatible with a clear understanding of correspondent's meaning." He expressly forbade "all expressions of opinion upon any matters; all political religious, and social biasses; and especially all *personal feelings* on any subject."[30]

Thus, by the date of Craig's circular, modern wire-service journalism had begun to take shape, characterized by centralized collection and distribution of news stories. News items gathered by telegraph had begun to take on a modern structure, marked by presentation of the most important facts first. The Civil War accelerated both these trends, confirming the NYAP as the leading (if not sole) distributor of official news from government departments and continuing the development of the inverted-pyramid structure.

The major catalyst for both of these trends was wartime government censorship of the telegraph. Instant reporting of battlefield events and government actions posed a new dilemma for the Federal government, between upholding the Constitution's guarantee of freedom of the press and censoring the news in order to maintain civilian morale and to deny the enemy useful information. Military and civilian officials gradually evolved a media strategy as the war progressed. Because they could not censor news at its point of distribution, the published newspaper, they attempted to control the flow of information from its sources, on battlefields and in telegraph offices.[31]

Official telegraphic censorship began in early July 1861 when the War Department barred all telegraphic dispatches concerning military operations. Secretary of War Simon Cameron placed officials of the American Telegraph Company in charge of censoring messages from Washington. During the first Battle of Bull Run on Sunday, 21 July, censors held back news of the northern defeat for a full day. Lawrence Gobright, Washington agent of the NYAP, recalled that the telegraphic censor had "permitted the *good* news to go, but suppressed the *bad*." Thus, the Monday editions of the Washington newspapers contained accurate and timely news of the North's defeat, but New York newspapers described the battle as a glorious Union victory. Since the War Department's censorship rules allowed Gobright to send any material already published in Washington newspapers, he transmitted an accurate account of the battle taken from a Washington newspaper, which appeared a day later in New York newspapers. When news of the defeat reached New York, the public greeted it with disbelief, many concluding that the news was a hoax perpetrated by speculators in order to manipulate the stock market.[32]

Soon after Bull Run, the State Department took over the job of censoring press dispatches. Samuel Wilkeson, Washington correspondent for the *New York Tribune*, later recalled that Secretary of State William Seward had vowed to "take the telegraph in my own hands. I will straighten out you men of the press." William Seward's son Frederick, the assistant secretary of state, oversaw censorship and appointed Harry Emmons Thayer, former manager of the American Telegraph Company's Philadelphia office, as the government censor. Thayer had wide powers to block all messages relating to "the civil or military operations of the government." Many reporters complained that Thayer had too much latitude, that he routinely censored dispatches that had nothing to do with military affairs and posed no danger to the war effort. As a result, Radical Republicans in Congress convened lengthy hearings on telegraphic censorship in January and February 1862. These unpublished House Judiciary Committee hearings, totaling some eight hundred manuscript pages, offer a unique window into Civil War journalism.[33]

After hearing testimony from nine newspaper correspondents, two telegraph operators, and Frederick Seward, the committee concluded that telegraphic censorship had gone too far. It made sense to censor important military information, but Seward and Thayer had overstepped their authority when they blocked news related to the civil operations of the government. The committee urged passage of a bill stipulating that the government should be allowed only to censor military information that might aid the Confederacy and that all other news should flow freely. Despite the committee's report, government officials continued to censor nonmilitary news items until the summer of 1866, including such trivial items as the departure of William Seward's nephew to Europe.[34]

Two other conclusions emerged from these hearings. To begin with, they showed how wartime censorship accelerated the trend toward the matter-of-fact, inverted-pyramid news story. A portion of NYAP correspondent Lawrence Gobright's testimony explains why the press association prized factual reporting:

> My business is merely to communicate facts. My instructions do not allow me to make any comments upon the facts which I communicate. My despatches are sent to papers of all manner of politics, and the editors say that they are able to make their own comments upon the facts which are sent to them. I therefore confine myself to what I consider legitimate news. I do not act as a politician belonging to any school, but try to be truthful and impartial. My despatches are merely dry matters of fact and detail.

More importantly, the Judiciary Committee found that the Lincoln administration had been deliberately using Gobright, and by extension the NYAP, as its official mouthpiece. Unlike previous presidents, Lincoln had not set up a pro-administration newspaper in Washington, preferring to disseminate his administration's views through the NYAP instead. The State Department's censorship policy reflected this, and Thayer permitted Gobright's dispatches to pass over the wires uncensored. Other correspondents either had to run the risk of having their dispatches blocked or heavily edited or had to mail them to their editors. According to Wilkeson, Gobright had "become within six months pretty much of a gov[ernmen]t institution. They make use of him for the public diffusion of intelligence of the war and the policy of the administration. . . . He has become almost as much the agent of the gov[ernmen]t as the agent of the newspapers." Wilkeson thought that the government used Gobright in this way in order to "vigorously sustain the Administration and to infuse vigor into the war." Wilkeson thought little of Gobright as a journalist, concluding, "They could not do that with me. I would not permit any department of the gov[ernmen]t to dictate a dispatch to the Tribune unless it coincided with my judgment." Gobright testified that he indeed obtained his news directly from government officers, often prepared in advance, and that none of his dispatches had been censored. He admitted that administration officials "have sometimes selected me as the medium for the general dissemination of information of public importance." Although Gobright testified that he did not object to telegraphic censorship, he complained bitterly of it in his memoirs published a few years after the war.[35]

Many of Gobright's dispatches survive in the War Department's records because Secretary of War Edwin Stanton ordered all USMT offices to retain the originals of all telegrams sent. These dispatches reveal the close and favored connection Gobright had with the War Department and the telegraph censors. Many of his telegrams were signed "Gobright per S.," possibly indicating that Frederick Seward or another government official had approved them for transmission to New York. Many were also marked "DH," meaning that they were sent "deadhead" or free of charge from the American Telegraph Company's Washington office. Gobright wrote some dispatches on War Department or USMT letterhead, indicating that he probably wrote them in those offices. Gobright earned these special privileges because his dispatches reflected the Lincoln administration's official line on military and governmental matters. For example, in late June 1862, rumors swirled that McClellan had suffered a major defeat in front of Richmond. On 29 June Gobright telegraphed NYAP's General Agent Daniel Craig, "As soon as the Department can obtain exact information . . . , it will be imparted to the public, whether good or bad. This dispatch is not intended for publication but for the information of the Press." On the following day, he telegraphed Craig that "nothing has been received to warrant the belief of any serious disaster. . . . Please have this bulletined as early as possible."[36]

Thus, Gobright and Craig exercised caution in their handling of rapidly changing war news. NYAP's procedures and caution later mitigated one of the war's biggest media disasters. On 19 May 1864, a story appeared in the *New York World* and *Journal of Commerce* (both Democratic newspapers hostile to the Lincoln administration) claiming that Lincoln had called up another 400,000 troops and had designated 26 May as a day of national humiliation, fasting, and prayer. However, no other New York newspaper published this story. The night editors of the *New York Times*, *Daily News*, and *New York Herald* immediately suspected that the dispatch was a forgery. Not only did it contradict the optimistic tone of news dispatches received earlier in the day, but it also arrived after the Washington NYAP office had sent its daily signal that it was closing for the night. Government officials quickly disavowed the story and a reporter for the *Brooklyn Eagle*, Joseph Howard, later confessed to planting it in order to manipulate the stock market. Few newspapers outside New York published this "Bogus Proclamation," likely only those who took non-NYAP dispatches over the lines of the recently formed Independent Telegraph Company. This incident demonstrated why the government placed a great deal of importance on controlling the flow of news during the war. It also strengthened the NYAP's reputation as the major purveyor of reliable news, even if much of it was officially sanctioned news.[37]

The wartime dispatches of Gobright and other correspondents showed how the structure of the news story was evolving from a traditional chronological account of events to the inverted-pyramid format that presented the most important facts first. Yet a full transformation was not complete until the 1870s and 1880s, some thirty to forty years after the rise of telegraphic newsgathering. This timing is consistent with technological and organizational developments in the telegraph industry and the newspaper press, particularly the company's decision to lease circuits to press associations and the close working relationship between Western Union and the WAP and NYAP after 1882.

This timing also helps explain why the increased use of the inverted-pyramid structure was not the same as a full embrace of the ideal of journalistic objectivity. Although Gobright characterized his "truthful and impartial" dispatches as "legitimate news" in 1862, reporters and newspaper editors did not come to regard objectivity—the decoupling of fact from opinion—as a professional norm until the late nineteenth or early twentieth century. In other words, the publication of factual, terse telegraphic dispatches was a necessary precondition for the later rise of journalistic objectivity, but the two are not coextensive.[38]

The transformation of the news story from Alfred Vail's transmission of Baltimore news to Samuel Morse in 1844 to the technical and organizational developments of the 1880s owed a great deal to the economic imperatives of the telegraph. The cost of telegraphic transmission was a key reason why correspondents sent the clipped, factual dispatches that were the heart of national and international news stories. However, as Civil War reporting showed, institutions that used the telegraph to collect and disseminate the news also played an important role. Wartime censorship of the telegraph forced correspondents to transmit inoffensive, stripped-down dispatches. Censorship also strengthened the NYAP's position as the dominant supplier of telegraphic news. Because the NYAP marketed its services to newspapers of all shades of opinion, its managers from Daniel Craig onward directed correspondents to provide only the facts, leaving editors to comment upon the facts as they saw fit.

Telegraphic Reporting and Changing Psychology of News Consumption

Three events reveal how the telegraph helped change the way Americans reacted to news of distant and important events. Observers first noticed a major shift in the psychology of news consumption in the early months of the Civil War. The Great Railroad Strike of the summer of 1877 showed how government officials used the telegraph as a communications tool to obtain real-time information on the progress of the strike. At the same time, newspaper coverage of the strike demonstrated that readers had become accustomed to sometimes jarring juxtapositions and geographic flattening of telegraphic news items. By the time of President Garfield's shooting and lingering demise in the summer of 1881, the public had become fully habituated to a modern sense of timeliness and newsworthiness.

Significant changes in how Americans understood themselves in relation to unfolding important events were apparent as soon as the first major battle of the Civil War. In late July 1861, Union Brigadier General Irvin McDowell advanced through northern Virginia, intending to take Richmond. Confederate forces met McDowell at Manassas or Bull Run, some twenty-five miles southwest of Washington. Because the closest telegraph station was at Fairfax Court House, several miles from the battle, mounted couriers carried news from the battlefield to Fairfax for transmission to Washington. A military telegrapher at Fairfax transmitted dispatches to the War Department every quarter hour. Despite early reports of Union success, this first major battle of the war was a Confederate victory. As quickly as the telegraphic dispatches arrived in Washington, they were posted and read aloud to an anxious crowd gathered in front of Willard's Hotel.

Such instantaneous news directly from the battlefield was a new phenomenon. On 22 July 1861, for example, the *New York Times* published a running account of the previous day's battle at Bull Run, reproducing some thirteen telegraphic bulletins sent between 11:00 AM and 5:45 PM. The newspaper's Washington correspondent commented on these bulletins that "the demand for intelligence is insatiable" and that the incoming dispatches were read to "vast crowds, who cheered vehemently, and seemed fairly intoxicated with joy." Similar crowds gathered in front of newspaper offices in New York from midafternoon until after midnight, and "each item of intelligence was canvassed and commented on." Similarly, the southern diarist Mary Boykin Chesnut, when McClellan was threatening Richmond in June 1862, noted that "we haunt the bulletin board" for telegraphic news.[39]

Chesnut mused in her diary, "When we read of the battles in India, in Italy, in the Crimea, what did we care? . . . Now you hear of a battle with a thrill and a shudder. . . . A telegram reaches you, and you leave it on your lap. You are pale with fright. . . . How many, many will this scrap of paper tell you have gone to their death?" Anna Shaw Curtis, sister of Captain Robert Gould Shaw of Massachusetts, wrote a friend in December 1862, "You cannot imagine the feeling of knowing that there is a battle going on. It makes us so nervous." The public excitement over battlefield news led Oliver Wendell Holmes to write an article in the September 1861 issue of the *Atlantic Monthly* about this new phenomenon of "*war fever*," one symptom of which was "a nervous restlessness . . . Men cannot think, or write, or attend to their ordinary business." One friend told Holmes that "he had fallen into such a state that he would read the same telegraphic dispatches over and over again in different papers, as if they were new, until he felt as if he were an idiot." In 1862 lithographer Louis Prang capitalized on this demand for instant war news by publishing a War Telegram Marking Map, measuring twenty-six by thirty-eight inches, that allowed civilians to follow troop movements "on the receipt of EVERY TELEGRAM from the seat of war" for twenty-five cents.[40]

In the summer of 1877 a nationwide railroad strike—the so-called Great Strike —paralyzed the country's industry and transportation for forty-five days. Violent street battles and demonstrations occurred in several cities around the country. In this atmosphere of unprecedented civil disorder, the telegraph proved important to the federal government's handling of the crisis. During the strike the army's Signal Service sent President Hayes reports every three hours (from 6:00 AM to midnight) from twelve to twenty cities daily. The Signal Service's major mission at the time was weather reporting and prediction, with observers trained in meteorology and telegraphy located in about two hundred cities and military posts. At times of national emergency, like the Great Strike, the Signal Service provided leaders in Washington with real-time intelligence. During the strike, enlisted Signal Service men dressed in civilian clothes, attended meetings and street demonstrations, and reported their findings to the War Department in Washington, who in turn passed the telegrams to President Hayes. As shown by his handwritten cabinet meeting notes, these dispatches proved essential to Hayes's decision making during the strike, particularly to what extent to use federal troops.[41]

The telegraph was also essential to newspaper coverage of the strike. This coverage displayed features of what sociologist Anthony Giddens calls the "collage effect," "the juxtaposition of stories and items which share nothing in common other than that they are 'timely' and consequential." On 20 July, for instance, the *Macon Telegraph and Messenger* gave about two columns of brief news items on the condition of the strike from various places under the general heading, "By Telegraph." Interspersed with and alongside of these items, the newspaper reproduced dispatches unrelated to the strike, including foreign news, the Signal Service weather forecast, and a statement of innocence by New York political boss William Marcy Tweed. Six days later, the *Chicago Inter Ocean* devoted three front-page columns to strike news sent by telegraph. Unlike its Macon, Georgia, counterpart, the *Inter Ocean* did not place unrelated stories in proximity to stories about the strike. However, it did not organize its coverage in any meaningful way, leaving it to the reader to determine the relative importance of the items.[42]

On 2 July 1881 Charles Guiteau shot President James Garfield at a Washington railroad station. Garfield lingered until mid-September, and telegraphic reporting of Garfield's condition held the nation enthralled during those two months. On the day of his shooting, anxious crowds gathered around the country at hotels and telegraph and newspaper offices. Updates on the president's condition arrived hourly, including details of his vital signs and bodily functions. Initial dispatches from Washington were reassuring; but at three o'clock dispatches announced that Garfield had hemorrhaged, sparking rumors that he had died. At nine o'clock the postmaster general sent a dispatch that was more reassuring, but at ten o'clock telegrams indicated that the president's death was imminent. The crowd at Broadway and Park Row, New York's newspaper district, alternated between hope and despair until the early morning. Within a few days, Garfield's condition stabilized and improved. As his doctors proclaimed him out of danger, the crowds diminished, and newspapers updated their stories daily instead of hourly. But in mid-August his condition again worsened. Hourly updates resumed and crowds once again gathered around the country in front of newspaper and telegraph offices.[43]

Instant news about Garfield's condition had a profound psychological effect. Crowds impatiently awaited the arrival of the latest news. Despite hourly updates, a *New York Times* reporter characterized the dispatches of 3 July as "exasperatingly slow and far between." In Brooklyn, noted another reporter, "crowds filled the streets" until early morning, "and every scrap of information . . . was eagerly sought for." On the following day, business activity and court proceedings stopped, as "men of all conditions jostled each other round the bulletins." In an extreme case, a widow living in Asbury Park, N.J., committed suicide on 6 July. She had been distraught over the death of a close relative, and the constant barrage of news about Garfield had so "excited and unnerved" her that "it was the one constant topic that engaged and absorbed her mind."[44]

Except for a few perceptive observers like Oliver Wendell Holmes, newspaper readers between Bull Run and Garfield's assassination seldom reflected on how their engagement with the news was changing. Newspaper editors, on the other hand, had to adapt themselves to the evolving demands of their readers, even if some were reluctant to do so. Colonel E. A. Calkins of the *Milwaukee News*, in the keynote address at the Wisconsin Editorial Association's 1874 meeting, commented on the changing psychology of news consumption:[45]

> The telegraphic dispatches must be published. Their importance, as matters of news, is a secondary consideration. People read them eagerly, whether they are important or not, because they came as they did, and from an association of ideas linked to the fact that the telegraph is employed for private messages only in vital emergencies. In telegraph matters, quantity takes precedence of quality. Not half the stuff which comes by telegraph would be printed in our columns if it came by any other mode of conveyance. But there is a public voracity for telegraphic news which will not be appeased by any substitute, nor by any assurance that what was omitted or did not come was of no account. This appetite grows on what it feeds on, like a novel-reading appetite in the young, or a whisky or opium appetite in the old. It is a peculiarity of American newspaper readers; and in Western newspaper readers it amounts to a disease or a mania. In our Milwaukee daily newspapers we publish each day more telegraphic matter than the *London Times* publishes in a week. The contents of our telegraphic columns exceed those of any other department of the newspaper. This feature is constantly expanding in dimensions. It grows like an excrescence upon journalism. We are helpless to resist it.

Newspaper coverage of the Battle of Bull Run in 1861, the Great Strike of 1877, and Garfield's death in 1881 demonstrated the validity of Calkins's remarks, that telegraphic reporting had changed the psychology of news consumption dramatically. Three important characteristics stand out. As the hunger for frequent news updates indicates, telegraphic reporting allowed Americans to participate in major events vicariously. The Civil War introduced Americans to what we today call the "news cycle," the expectation of receiving frequent and timely updates on evolving news stories. And, as Oliver Wendell Holmes astutely observed in 1861, telegraphic reporting created a psychological demand for news that had not existed with the same intensity or urgency beforehand. Garfield's assassination revealed most fully how the public's expectations of timeliness and newsworthiness had changed in the intervening twenty years.[46]

Civil War battles and Garfield's lingering death revealed that news consumption in the late nineteenth century retained its public character. Historians have long noted that Americans of the eighteenth and early nineteenth centuries got their news in communal settings like churches, coffeehouses, merchants' exchanges, and post offices, and that the growth of mass-circulation newspapers after the 1830s shifted news consumption from the public to the private sphere, a process largely complete today.[47] Yet major news events caused crowds eager for the latest updates to gather at newspaper and telegraph offices and especially at hotels. Hotel proprietors posted the latest telegraphic news on bulletin boards to draw the curious into their saloons and restaurants. Hotels were therefore central to the sociology and geography of information transmission. In 1865, for example, three of the five Washington, D.C., offices of the American Telegraph Company were located in hotels. New York's Astor House hotel had its own telegram message blanks that noted its connection to the lines of three telegraph companies. Within two hours of Garfield's shooting, according to a *New York Times* reporter, nearly every New Yorker had heard the news: "The telegraph carried it to all the principal hotels, and from these common centres of information it radiated to the smallest tenement-house districts. . . . The newspapers were receiving dispatches every few minutes, and as fast as they came from Washington they were posted on the bulletin boards."[48]

Telegraphic reporting of the Great Strike of 1877 showed how wire-service journalism had shifted the task of determining the relative importance of news stories from editors to readers. In 1865 James Russell Lowell had this effect in mind when he criticized telegraphic reporting for failing to convey "the true proportions of things. . . . In brevity and cynicism it is . . . as impartial a leveler as death. . . . In artless irony the telegraph . . . 'confounds all distinctions of great and little.'" Rollo Ogden, writing at the turn of the century, extended Lowell's critique to overseas news, complaining that the cables had "thrown everything out of perspective." Foreign news by cable had become "an undifferentiated pulp," a "huge mass of undigested information (mostly *mis*information) and opinion flung at our heads." Exhaustive content analysis of daily newspapers has shown a similar, if less jaundiced conclusion, that telegraphic reporting indeed separated facts from their contexts. The geographic and temporal sweep of events like the Civil War or the Great Strike demanded a haphazard presentation of telegraphic bulletins on newspaper pages.[49]

Technological and institutional underpinnings facilitated these changes. By 1861 telegraph lines blanketed the entire country, and press associations and telegraph companies had established good working relationships that sped the flow of news. By 1881, Western Union and press associations had come to work so closely together that they controlled the flow of the nation's news. If these three news events fully revealed the changed psychology of news consumption, they also exposed the institutional and technological foundations that would characterize the collection and distribution of news well into the next century.

What Happens in the Telegraph Office
Stays in the Telegraph Office

Today, widespread texting, emailing, and instant messaging threaten to lead to the demise of the grammatical sentence and the deterioration of writing skills. Many educators are concerned with their effect on the quality of writing by young Americans. However, some note that teens engage in code switching, distinguishing between the writing done for school and that for text messaging, email, and social networking sites. Unlike these concerns about twenty-first-century electronic communication, nineteenth-century observers were much more optimistic about the telegraph's effect on writing. Yet similar conclusions might apply with equal force to the telegraph's relationship to writing.[50]

Commentators who welcomed the telegraph's effect on writing in the nineteenth century and today's critics of electronic communications share the belief that new communication media have significant and lasting effects on prose composition. After all, these technologies interact strongly with the written word. They have changed how Americans communicate with each other and produce and consume news. Yet dissent from these views suggests that the relationship between technology and the written word is more complex and ambiguous than these hopes or fears.

Both the telegraph and electricity served as metaphors and plot devices in nineteenth century fiction, poetry, and essays.[51] However, the telegraph's effect on prose style itself is far less obvious. It is clear that American literary writing became leaner, more direct, and more forceful sometime after midcentury, a process Edmund Wilson has called "the chastening of American prose style." Wilson attributes the changes in American prose to a complex combination of factors: the faster pace of American life after midcentury, the spread of mechanization during and after the Civil War, the influence of newspaper style upon literature, and the popular writing and addresses of unpretentious westerners like Lincoln and Twain. Edgar Allan Poe found that he could earn a living as a writer only if he wrote short fiction for the expanding magazine market, which demanded "the curt, the well timed, the terse." Contemporaries praised the Great Triumvirate of American political oratory, Clay, Webster, and Calhoun, for their "blunt earnestness," "simplicity," and "plain, terse, clear, and forcible" expression. As several historians have pointed out, such direct and forceful prose and oratory was central to the nationalist project of forging a unique American idiom and literature in contradistinction to aristocratic and bloated British English.[52]

The telegraph's role in this process was ambiguous, indirect, and likely less important than political and cultural factors. While the technical and economic constraints of telegraphy encouraged customers and operators to condense telegrams as much as possible, it is hard to see how the new technology might have changed other forms of prose composition. After all, few Americans regularly sent or received telegrams. Furthermore, telegraphy was a specialized form of communication, best suited to the urgent transmission of factual information, while literary prose is produced and consumed at a more leisurely and deliberate pace.

Most Americans had contact with telegraphic language through their reading of newspapers. But the scope and amount of telegraphic news reports were limited. The *New York Herald* boasted that it printed thirty-six and a quarter columns of telegraphed news during the first two weeks of January 1848. Of this volume, summaries of legislative proceedings from Washington, Albany, and Harrisburg occupied about 74 percent of the content; and market reports and shipping news took up another 11 percent. Foreign news, southern and western news, and miscellaneous items accounted for the remaining 15 percent of the column space. Most early NYAP dispatches were of little literary value, consisting of legislative summaries, produce market quotations, and shipping news. Furthermore, less than 10 percent of the news stories appearing between 1847 and 1860 were telegraph dispatches, and these tended to be items containing hard, quantitative information.[53] Although the volume of wire-service dispatches increased dramatically during and after the Civil War, it is hard to see how telegraphed news stories brought about a thoroughgoing revolution in prose composition.

Examples of literary styles influenced by the telegraph are rare and ambiguous. Some literary scholars have described Emily Dickinson's poetry as "telegraphic," noting in particular the linguistic compression, erratic punctuation, and uneven tempo of her poems. However, little evidence exists that she deliberately emulated the structure and format of telegrams. The resemblance between her poems and telegrams was probably coincidental, owing to the rapidity of composition in both cases. Similarly, in 1940 Ernest Hemingway credited the style guide of the *Kansas City Star*, the first newspaper he worked for as an eighteen-year-old cub reporter in 1918, for imparting "the best rules I ever learned for the business of writing." The *Star*'s style guide directed reporters: "Use short sentences. Use short first paragraphs. Use vigorous English. Be positive, not negative." Hemingway's distinctive voice probably owed more to the *Star*'s style guide than it did to his experience as a cable correspondent. Hemingway's fellow cable correspondent

William Shirer later recalled that "cablese" "made for stilted style, a barrenness of language," poorly suited for prose composition generally.[54]

The telegraph was therefore part of a constellation of political and cultural factors that condensed and streamlined prose in the nineteenth century. Why then did contemporaries and later commentators assign it a large role in this process? Before the Civil War, Americans credited the telegraph with much more linguistic influence than it actually exerted because telegraphic language reinforced traits which they saw, or wished to see, in the American character. The telegraph merged two elements, mastery of technology and democratic notions of language, which lay at the core of the drive to forge a national identity.[55] Telegraphy's symbolic power allowed Americans to see themselves and their idiom as they wanted—united, progressive, and democratic.

The program to construct democratic and distinctively American forms of expression contained three broad elements. Politicians and popular orators in this period strove for a blunt and direct style. They believed that unadorned and forcible speech best expressed the progressive and egalitarian spirit of the age. Politicians favored concise forms of public speaking, such as Lincoln's brief yet poignant Gettysburg Address. Familiarity in address, marked by such greetings as "friend" or "fellow," was another way in which Americans leveled distinctions of wealth and education. Telegraphic communication, which dispensed with the flowery greetings and closings of contemporary letters, fit well with this informality. Finally, Americans embraced a shift in literary style, a move away from more formal cadences toward simpler sentence constructions and shorter, more powerful words. "Guts," noted Emerson, "is certainly stronger than intestines."[56]

Many Americans in the middle third of the century hoped to establish a uniquely American literature based on these three elements. The editor Rufus Wilmot Griswold wrote in 1854 that authors on this side of the Atlantic "have been so fearful of nothing else as of an *Americanism*, in thought or expression." In its first issue in 1837, the *Democratic Review* found it tragic that "we have no national literature." American writers, it charged, were "enslaved to the past and present literature of England." The "anti-democratic character" of such slavish imitation was "poisoning at the spring the young mind of our people." The new republic required a new literature. The task was immense: "All history has to be re-written; political science and the whole scope of all moral truth have to be considered and illustrated in the light of the democratic principle."[57] Telegraphic brevity was particularly suited to this task.

The Civil War largely settled these questions of national identity and unity, removing much of the impetus to forge a distinctive American language and literature. After the war, Americans heralded the entrance of their country onto the world stage as a commercial, if not yet a military and political, power. The completion of the Atlantic cable in 1866, a joint British-American project, promised to make English the global language of commerce. Commentators no longer sought to construct an American idiom in contradistinction to British English but looked forward to the commercial and political dominance of the Anglo-Saxon race and tongue. For example, a contributor to the *Journal of the Telegraph* predicted in 1869 that within a hundred years the global businessman would need no other language. The "compactness" and "expressiveness" of English, the "king of dialects," made it better suited for cable traffic and commerce than any other language. Another writer in the same journal predicted that English would become the globe's dominant language within two hundred years, because no other language was "so cheap and expressive by telegraph."[58]

In an important sense the predictions of nineteenth-century commentators have come to pass. American prose did become leaner and more streamlined after the Civil War, and English has become the global language of commerce and culture. However, the role of the telegraph in these transformations was indirect, ambiguous, and likely less important than other cultural and political factors. Although the telegraph, working through press associations, played a key role in changing the structure and tone of the newspaper story, little evidence exists to show that telegraphy changed prose composition more generally. The uncertain and muted effects of the telegraph upon literary style suggest that a technology's effects upon cultural production depend on the scope and intensity of its use. In the telegraph's case, the medium was only part of the message.

U.S. Military Telegraph Corps linemen stringing wire in camp. From Alexander Gardner, *Gardner's Photographic Sketch Book of the War* (Washington, D.C.: Philip & Solomons, 1866), plate 62. Courtesy of Rare Book and Manuscript Division, Cornell University Library, Ithaca, New York.

This patent drawing of George Beardslee's field telegraph shows that he designed it for ease of use by untrained operators in the field. Chief Signal Officer Albert Myer hoped to use the Beardslee instrument to place electric telegraphy under his military authority. Secretary of War Edwin Stanton overruled Myer and left telegraphing to the U.S. Military Telegraph Corps, staffed by civilian operators and linemen. Courtesy of the National Archives, at www.ourarchives.wikispaces.net/, accessed 10 Dec. 2011.

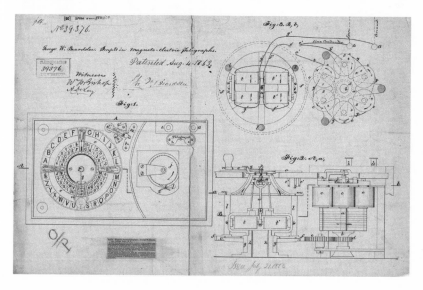

This Joseph Keppler cartoon caricaturing Jay Gould and William H. Vanderbilt captured the antimonopoly sentiment directed at Western Union after Gould's 1881 takeover. *Puck,* 23 Feb. 1881.

Frequent updates on Garfield's condition posted on Broadway, New York. *Frank Leslie's Illustrated Newspaper,* 3 Sept. 1881.

This photograph of New Street in lower Manhattan, taken after a March 1888 blizzard and ice storm, gives a sense of the magnitude of the communications infrastructure on which New York's financial markets depended. Courtesy of New York Stock Exchange Archives, NYSE Group.

New York Stock Exchange floor traders closely inspect the "ticker tape" in 1899. Courtesy of New York Stock Exchange Archives, NYSE Group.

The main New York office of bucket shop chain Haight & Freese, showing the proximity of the tickers to the order desk. From *Haight & Freese's Guide to Investors* (Philadelphia, 1899).

The cover of Postal Telegraph Company's employee magazine, January 1919. From entry 38, RG 28, National Archives, Washington, D.C.

Advertising pamphlet in the 1950s for Western Union's Desk-Fax. From the collection of Robert Harris, West Sand Lake, New York.

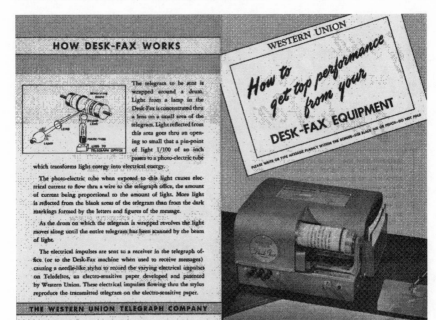

CHAPTER FOUR

"The Ticker Is Always a Treacherous Servant"

THE TELEGRAPH AND THE RISE OF MODERN FINANCE CAPITALISM

In his official history of the New York Stock Exchange, Edmund Clarence Sted-man, the celebrated "broker-poet" of Wall Street, described the multitude of tele-graph and telephone wires that connected the Stock Exchange to the rest of the country in the 1880s. He recalled, "No bird could fly through their network, a man could almost walk upon them; in fact, they darkened the street and the windows below their level." For Stedman this sight drove home "the meaning of the Stock Exchange as the focus to which all currents of American purpose and energy converge" and provided an "ocular demonstration of the relations of the New York Exchange to the Republic in its entirety, and even to the world overseas."[1]

The flow of information has always been essential—if at times less visible—to the operation of financial markets. In the early nineteenth century, the British banking family, the Rothschilds, set up a far-flung communications network that gave them a substantial time advantage over their commercial rivals. Around 1840, William C. Briggs, a Philadelphia stockbroker, set up a private semaphore telegraph between New York and Philadelphia to conduct arbitrage in stocks listed on both exchanges. Briggs abandoned his line in 1846 after a Morse tele-graph line connected both cities.[2] The electromagnetic telegraph not only out-paced his optical system but, more importantly, ended his monopoly over the rapid transmission of stock prices. Thus, the electromagnetic telegraph rendered financial information public, available to anyone willing to pay for a message. Stock and commodity prices from major commercial centers became staples of wire-service journalism and permanent fixtures in daily newspapers.

After the Civil War two inventions transformed both the telegraph industry and the nation's financial markets. In 1867 Edward Calahan invented the ticker, a cheap and reliable printing telegraph that broadcast real-time financial information from exchange floors to anyone subscribing to the service. Within a few years, the market for ticker service expanded dramatically, with hotels, saloons, and bucket shops (betting parlors in which patrons wagered on price movements of stocks and commodities) renting machines in addition to brokers and bankers. The second invention was Thomas Edison's quadruplex, a device that allowed four messages to be sent simultaneously over one telegraph wire. Western Union acquired ownership of the quadruplex in the mid-1870s, effectively quadrupling the bandwidth of circuits on which it was installed. As a result, after about 1880 Western Union's managers reversed their long-standing reluctance to leasing circuits to press associations and stockbrokers. By the turn of the century, nearly all brokers leased private circuits as well as tickers to transmit orders and quotations between branch offices and exchange floors. Together, the ticker and leased-wire networks formed the backbone of modern American finance capitalism.[3]

Until the spread of the ticker in the 1870s, financial markets remained much as they had been since their origins in early modern Europe—markets in tangible goods. Stock and commodity traders met "on 'Change'" during set hours to deal stock and bond shares or warehouse receipts of agricultural produce. For example, until the advent of continuous trading in 1871, brokers on the New York Stock Exchange traded shares twice a day, at 10:30 AM and 1:00 PM. Sellers gave their shares to the president, who then auctioned them off to the highest bidder. Information was of course crucial to these transactions, made manifest by the hordes of runners carrying orders and completed trades between brokers' offices and exchange floors. With the advent of continuous trading and the spread of the ticker in the 1870s, however, the nature of financial markets changed. Instead of markets that relied on information, they increasingly became markets *in* information. Participants came to regard markets less as places to trade tangible things and more as the flow of quotations printed by the ticker and posted on distant blackboards.[4]

Technological innovation, first the electromagnetic telegraph after about 1850 and then the ticker and quadruplex after about 1870, provided the foundation for the development of modern finance capitalism during the last third of the nineteenth century. Information from the records of major stock and commodity exchanges and telegraph companies reveals how and to what extent telegraph technology changed the structure and nature of financial markets during this period. Telegraph technology, particularly the ticker, recast the psychological and geographic relationships between market participants and markets. As financial markets increasingly became markets in information, control of and access to the flows of quotations became a major source of conflict between exchanges, telegraph companies, brokers, and bucket shops. By broadcasting quotations to a wider and wider audience, the ticker and telegraph network enabled the dramatic growth in stock trading and ownership in the twentieth century.

The Business and Technology of Financial News Reporting

The publication of financial news, mainly currency exchange rates and commodity prices, dates back to sixteenth-century Europe. From the very beginning there existed a close connection between business information and newspapers. The publication of "prices current" marked the start of the modern newspaper in Europe and the American colonies. The publication and dissemination of such financial information were important reasons why businessmen in many American cities formed merchants' exchanges and chambers of commerce in the first half of the nineteenth century. Indeed, the transmission of commodity prices from important market centers was a mainstay of the New York Associated Press in its early years before the Civil War.[5]

While the telegraph transformed and modernized many sectors of the American economy in the nineteenth century, particularly the nation's financial markets, this transformation was quite different from what early observers had expected. After its introduction in the 1840s, most businessmen anticipated that the telegraph would reduce opportunities for speculation and manipulation by broadcasting market information rapidly and widely. A writer in *DeBow's Review* in 1854 thought that the telegraph had reduced "cotton and stock gambling" by 95 percent, from $40 million a year to only $2 million. As the demise of William Briggs's private optical telegraph line showed, these expectations proved largely correct. The early telegraph network reduced opportunities for speculation and arbitrage, both of which arise from informational asymmetry and uncertainty.[6] However, in 1890 Western Union's president Norvin Green testified to Congress that 46 percent of his company's message traffic was "purely speculative," including "stock-jobbing, wheat deals in futures, cotton deals in futures," and horse racing odds, while only 34 percent pertained to what he considered "legitimate trade."[7]

A closer examination of two related developments helps to reconcile these disparate assessments and shows how the telegraph network came to abet speculation. Modern American financial markets originated as an outgrowth of speculation in government-issued bonds and paper currency during the Civil War.[8] The range and depth of these transactions increased dramatically during and after the war, facilitated by telegraphic reporting of financial and general news. After the Civil War, the ticker and later the quadruplex deepened the connection between speculation and telegraphy by increasing the quantity and accelerating the flow of information from exchange floors to market participants. By 1880, the organizational and technological infrastructure of modern financial markets was firmly in place.

The Civil War sparked a dramatic increase in speculation in stocks, government bonds, gold, and agricultural commodities. Financial writer James K. Medberry observed a few years after the war that "the whole population of the North gave itself up to a speculative frenzy.... The war made us a great people, made us also a nation in whom speculative ideas are predominant." The rampant speculation in gold—or, more properly, federally issued paper currency—during the war is well known. The so-called Bogus Proclamation incident in May 1864 demonstrated the power of telegraphic information (even false information) to influence the gold market. Also, during the war more and more people speculated in stocks. For example, Western Union issued a staggering series of dividends that more than tripled its nominal capitalization between September 1861 and May 1864, sparking a speculative frenzy in its stock. Western Union executive James D. Reid recalled that residents of Rochester, the company's headquarters, sold off "pianos, guitars, furniture of various kinds, mortgages, and homesteads" to buy Western Union shares. "While the public mind was thus excited," Western Union's president Hiram Sibley decided to declare a 100 percent dividend, doubling its capital to the huge sum of $21 million. These unprecedented dividends created a furious demand for shares.

Among those caught up in this speculative frenzy were telegraph managers and operators, not surprising given the access they had to war and financial news ahead of the rest of the public. Anson Stager, serving as both U.S. Military Telegraph Corps (USMT) superintendent and Western Union superintendent, and George Ladd, Western Union's California superintendent, both made fortunes leveraging their advance knowledge of war news to speculate in gold. Some operators as well as managers also attempted to use advance knowledge of the news for their financial benefit. Patrick Mullarkey, a USMT operator at Louisville with an otherwise exemplary record, tried to entice David Homer Bates of the USMT's Washington War Department office into transmitting important military and political news to Louisville gold speculators. Bates indignantly rejected Mullarkey's proposition, despite the lure of $50,000. It is unknown if Mullarkey ever consummated his scheme, but Bates and Ohio telegrapher George Kennan contented themselves with buying Western Union shares. Several southern

operators also engaged in cotton and stock speculation on the basis of their advance knowledge of the arrival of ships that made it through the northern block-
ade. Thus, purchase of telegraph stocks or speculation on the basis of war news
seemed to be common among operators and managers.[9]

The ability to monitor real-time trades on exchange floors from a distance
originated with the gold indicator, invented in 1866 by Samuel Spahr Laws, a
founder and vice-president of the New York Gold Exchange. At first, Laws's device indicated the price of gold trades by means of a large double-faced dial indicator, one face visible to traders on the floor of the Gold Exchange, the other
to spectators on the street outside. Later in 1866 Laws devised a small indicator
suitable for remote operation. It had three wheels, two displaying numerals and
a third displaying fractions. Laws resigned from the Gold Exchange, started his
own firm called the Gold and Stock Reporting Company, and entered into a contract with the electrical manufacturing firm of Charles T. and J. N. Chester to
supply instruments. To transmit gold quotations, Laws employed an operator
at the Gold Exchange to send the trades using a keyboard transmitter with keys
that sent electrical impulses, the length of which corresponded to the several
numerals and fractions needed to indicate the trades at customers' instruments.
By the end of 1866, Laws had about fifty customers for his gold indicators. Laws
envisioned printing out the results of gold trades onto a paper tape, using methods similar to those used in printing telegraphs like those designed by Royal E.
House and David Hughes. Two skilled electricians, Franklin L. Pope and Thomas
Edison, refined Laws's design and gave him a commercially practical printing
instrument. By the end of 1868 the Gold and Stock Reporting Company had
installed some three hundred instruments in New York's financial district. Laws
entered into an arrangement with a small telegraph line, the Bankers and Brokers
Telegraph Company, to supply New York gold quotations to about thirty offices
in Philadelphia, with plans to expand the business to Boston. According to Western Union executive James D. Reid, the performance of Laws's Philadelphia circuit was extraordinary, with results of gold trades printed a mere fifteen seconds
after they were made on the Gold Exchange floor.

In August 1869 Laws sold his company to a rival, the Gold and Stock Telegraph Company (G&S). Managers of both companies worried that their respective ticker patents might interfere with each other, so a merger seemed the wisest
course. G&S had its origins with Edward A. Calahan, who had been a telegraph
operator and electrician since 1850. In 1867 Calahan invented a printing telegraph that could rapidly print letters as well as numbers and fractions, thus making it suitable for stock trades as well as gold transactions. To exploit his inven-

tion, he organized G&S in September 1867, although another six months elapsed before Calahan perfected his instrument and obtained a patent for it. The new G&S began soliciting subscriptions from brokers and bankers at $6 a week plus a one-time installation fee of $100.

Like all new technologies, the ticker required technical refinement before it gained widespread acceptance. Subscribers, after all, demanded reliable and maintenance-free operation. Laws, Calahan, Edison, Pope, and other electricians wrestled with three major challenges that involved both the ticker and its transmitter. While the term *ticker* typically refers to the printing telegraph that prints stock quotations onto a paper tape, the transmitter was an equally important part of the system. Both Laws and Calahan worked for several months to get transmitters that could reliably send stock and gold quotations to many tickers simultaneously. A related problem was that the transmitter and a customer's ticker often fell "out of unison." The printer inside the ticker printed out letters or numbers based on the duration of an electrical pulse sent by the transmitter. If the type wheel of the ticker was not in the correct position at the beginning of the pulse, it would print a letter or number other than the one desired, resulting in a string of gibberish. This required a visit from a G&S electrician to reset the ticker manually. Henry Van Hoevenburgh devised an automatic unison in 1871 to prevent this. Aesthetic as well as technical considerations were important as well. Calahan's original design required a local battery at each ticker. Contemporary telegraph batteries were essentially zinc and copper electrodes placed inside acid-filled glass jars, and they needed their acid replenished twice a week to operate properly. Ticker customers were understandably reluctant to risk acid spills that damaged carpets and furniture. By 1869 Edison and Pope had perfected designs that obviated the need for a local battery.[10]

The collection and transmission of ticker quotations were major technical and organizational undertakings. To take the New York Stock Exchange as an example, trading began at ten o'clock with traders in particular stocks gathering at different parts of the exchange floor. The exchange allowed G&S to install telegraph keys at several points on the perimeter of the trading floor. G&S sent reporters onto the floor to shadow each of these groups of traders. Each reporter was a trained telegraph operator. A G&S reporter recorded the results of trades made in the stocks he monitored, ran to one of the telegraph keys on the wall of the trading room, and telegraphed the transactions to Western Union's central office. To save time, reporters gave stock names standard one to three letter abbreviations (which we retain today in stock quotations). At Western Union's central office, a receiving operator recorded the quotations. Another operator seated

TABLE 4.1.
Significant Ticker Designs, 1866–1929

Ticker	Inventor	Year	Comments
Calahan 3-Wire	Edward A. Calahan	1867	
Universal 2-Wire	Thomas A. Edison	1870	Under contract with G&S
Phelps 2-Wire	George M. Phelps	1870	Western Union electrician
Manhattan 1-Wire	Charles T. Chester	1871	Developed by Manhattan Quotation Co., a G&S rival
Phelps 1-Wire	Henry Van Hoevenburgh,	1874	Adapted from Phelps 2-Wire
	George M. Phelps	1876	Weight or escapement driven
New York Quotation 2-Wire	Stephen D. Field	1880	Developed by Commercial Telegram Co., a G&S rival
Scott 2-Wire	George B. Scott	1883	Western Union electrician
Scott 1-Wire	George B. Scott	1885	
Burry Page Printer	John Burry	1890	Developed by Stock Quotation Telegraph Co., printed on 5½-inch-wide roll and used mainly by news agencies
Scott-Phelps-Barclay Page Printer	George B. Scott, George M. Phelps, J. C. Barclay	1903	Page printer adapted from Scott 2-Wire
Western Union Automatic Self-Winding	Western Union engineers	1923	
Ticker 5-A	Joint project of Western Union, NYSE, and Teletype Corp.	1929	High-speed, 300 characters/minute

SOURCES: From "A General Survey of Ticker Service Furnished by the Western Union Telegraph Company," Oct. 1929, WUTC; *Cyclopedia of Applied Electricity*, vol. 7 (Chicago: American Technical Society, 1919).

next to him then sent the quotations over the ticker transmitter. The transmitter, based on previous designs for printing telegraph transmitters (such as the House and Hughes telegraphs), operated by means of a keyboard. The operator pressed keys corresponding to letters and numbers to generate the quotations that appeared on ticker tapes. In 1879 some 120 circuits, each with several tickers on them, radiated from the keyboard transmitter. Each circuit required three wires, one to control the letter type wheel, the second to control the number type wheel, and the third to send the signal to print the desired character.[11]

The Gold and Stock Telegraph Company proved to be lucrative almost immediately. From its start in October 1867 with twenty-five subscribers, it grew to two hundred customers by the end of 1868 and planned to build its own factory to make instruments. More significantly, by the start of 1869 the company had also tied up with the Pacific and Atlantic Telegraph Company, a rival of Western Union's, to supply New York stock and gold quotations to subscribers in seven midwestern and southern cities. Western Union, seeking to stave off competition from firms such as the Pacific and Atlantic, began a Commercial News Department (CND). CND was a joint venture with the New York Associated Press to provide newspapers and other subscribers with general and financial news at several times during the business day. The CND service did not use tickers but was more akin to the distribution of press dispatches. Western Union and G&S also agreed not to compete with each other: G&S was to supply quotations only in New York, and CND everywhere else. This agreement was easy to reach: G&S president Marshall Lefferts was also a Western Union vice-president.[12]

Western Union's president William Orton had "grave doubts" about the CND business. He admitted on several occasions that it was not very profitable but that it was essential to stave off competition from rival telegraph companies. A more pressing concern was legal liability for incorrect transmission of quotations. In 1869 several cotton brokers in New Orleans sued Western Union for trading losses resulting from erroneous quotations. The company's lawyers advised Orton that there was "no successful defense" against this liability, leading him and other managers to consider abandoning the CND business altogether.[13]

However, Orton knew that financial news reporting held major advantages for Western Union. It was, of course, lucrative. While Western Union's CND was not very profitable, G&S had amply demonstrated that ticker service was. One reason was the different marginal cost structure of ticker service. Transmission of ordinary message telegrams was labor intensive, requiring a sending and receiving operator. Additional ticker subscribers, on the other hand, could be accommodated with little or no increased labor costs. In addition, quotations generated commercial telegram business. Furthermore, the ticker market could be easily dominated, if not monopolized, through ownership of patent rights and exclusive agreements with stock and commodity exchanges. Most importantly, ticker service was not subject to the threat emanating from the strong postal telegraph movement. Had the government opted to nationalize the telegraph industry under the provisions of the National Telegraph Act of 1866 or to set up a competing postal telegraph, Western Union would be able to remain in business as a purveyor of ticker quotations. Indeed, Orton's two most pressing concerns around 1870 were the threat of renewed telegraph competition and the specter of a postal telegraph bill passing Congress. A Western Union ticker service would insulate the company from the worst eventualities of a postal telegraph or strong competition in the message telegram market, while still generating high profits. In November 1870 Orton wrote Anson Stager, Western Union's Chicago superintendent, that "sending messages between New York and Chicago is no longer a monopoly." Instead, he thought that the "great future of telegraphy" lay in "the distribution of markets among customers in the large cities." Two months later he confided to Stager that he feared that passage of a postal telegraph bill was inevitable and that Western Union ought to prepare for a possible government takeover of the message telegram business. This would leave Western Union with the business of the "distribution of markets between stock exchanges and boards of trade," a market "susceptible of almost infinite expansion" and not subject to government control or regulation.[14]

At the time Orton wrote these letters to Stager, he was negotiating with Lefferts to acquire G&S. In May 1871 the two sides reached agreement. G&S doubled its existing capital from $1.25 million to $2.5 million and gave half of the new shares to Western Union. Western Union thus became half owner and de facto controller of G&S, but G&S still maintained a nominally separate corporate existence. Western Union also insisted that G&S take over the CND service because it feared legal liability over the incorrect transmission of quotations. Also, if a successful postal telegraph bill stripped Western Union of its message telegram business, it would still maintain half ownership of the lucrative G&S. As Orton described the division of markets between Western Union and G&S to Lefferts in 1872, "The Western Union will receive messages from anybody to anybody. The G&S will provide instruments for delivering reports destined only for subscribers, and will also establish private telegraphs for private purposes."[15]

This latter business, private-line telegraphs, was an offshoot of the company's financial news business. This business supplied private parties like banks, mercantile houses, and manufacturing firms with their own telegraph lines and instruments. These instruments were simple to operate and did not require a trained telegrapher. The instrument consisted of a keyboard transmitter and a printer similar to a stock ticker for receiving messages. Private-line telegraphs were forerunners of the telephone in their social and business applications and in their use of a switchboard to connect two subscribers with each other for direct communication. Typical uses of private-line telegraphs were to connect a factory with its head office, to link together banks and clearinghouses, and to connect New York law firms to the courthouse.[16]

Some twenty-five national banks of New York used private-line printers connected through a central switchboard at the clearinghouse from 1869 until superseded by the telephone in 1879. By allowing subscribers to conduct transactions with each other and to balance their accounts more quickly and accurately, this network transformed the operations of the largest banks that sat atop the nation's financial pyramid. G. L. Wiley, an electrician with the Gold and Stock Telegraph Company, recalled that, before the system's adoption, bank cashiers "waited impatiently" for transactions to clear and were "often in doubt as to whether the balance were debit or credit" until their messengers returned with the completed transaction slips. Sometimes "the clearing house failed to balance for hours." After 1869, however, transactions and reconciliations occurred within minutes.[17]

A related use, the so-called domestic or district telegraph, was intended for the home and had a simple dial with six or seven settings whereby a customer could summon the police, fire department, doctor, or telegraph messenger boy. The private-line business originated in 1869 when Franklin L. Pope, James N. Ashley, and Thomas Edison formed the American Printing Telegraph Company to commercialize an easy-to-use design devised by Pope and Edison. The company met with limited success and attracted the attention of Western Union, which soon bought it out. When Western Union assumed control of G&S in 1871, it turned over the private-line business to it. G&S charged $30 a month for a two-mile line with two instruments, with an extra charge of $5 per month for each additional mile or $15 a month for each additional instrument. The private-line business grew rapidly in the 1870s, with subscribers in New York and a dozen other cities. The business peaked in 1876 and 1877 with revenues of about $1 million and profits about $90,000. The rapid commercialization of the telephone after 1877 caused the steep decline of the private-line market, and in 1881 G&S sold it off to various local Bell operating companies.[18]

Thus, by the early 1870s the ticker had become a mature technology that gave subscribers rapid, reliable, and maintenance-free printing of stock and commodity quotations. Within a decade of the commercial introduction of the ticker, Western Union had constructed an elaborate communications infrastructure for the country's financial markets. By 1877 the company provided direct telegraphic communication between New York's stock, cotton, and produce markets and their sister markets in some fifteen cities. The company's office on the New York Stock Exchange floor, for example, handled more than two thousand messages a day. In addition, some eighty-five New York hotels and restaurants rented instruments that printed general and financial news items. Western Union also supplied brokers and bankers the prices of U.S. and European government securities traded in London, Paris, and Frankfurt. Performance was extraordinary: in 1883 average transmission times for messages between the New York and Boston stock exchanges were less than a minute, between the Chicago and New York produce exchanges under two minutes, and between the New York and New Orleans cotton exchanges about five minutes. By 1886 G&S rented more than twenty-two hundred tickers and provided ten different kinds of quotations to subscribers. By 1908 the company had just under three thousand tickers in service around the country, with more than one thousand in New York and nearly five hundred in

TABLE 4.2.
*Gold and Stock Telegraph Company
Instruments in Service*

Year	Tickers	Private Line Printers
1867	25	na
1868	192	na
1869	572[a]	na
1870	741	na
1871	830	na
1872	~1,000	~500
1873	1,324	483
1874	1,774	654
1875	1,400	612
1876	1,310	692
1877	1,398	817
1878	1,342	~720

SOURCE: Board of Directors Minutes, Gold and Stock
Telegraph Co., WUTC
    Note: About 60 percent of the company's ticker rentals at
this time were in New York. After Western Union assumed
control of G&S in 1871, G&S operated its private-line business.
    [a]Of this increase, 203 were due to absorption of Laws's
company.

Chicago. Only seventy-six tickers were in service in the South and none on the
Pacific coast. Indeed, according to the Bureau of Labor Statistics, NYSE ticker
service was not available west of Kansas City until 1926. This geographic expansion of NYSE ticker service and the introduction of a new high-speed ticker in
1930 were key factors in the New York exchange's growing dominance of the
American securities market.[19]

Despite Orton's hope that Western Union's financial news services would be
insulated from antimonopoly agitation, the company's dominance of this market
attracted critics' attention. They feared that Western Union's quotation and financial news services might be used to help insiders—whether company employees
or favored customers—to manipulate markets. Benjamin Butler, for example,
made this fear a cornerstone of his 1875 House Judiciary Committee report calling for the nationalization of the telegraph: "There is no industry, no interchange
of commodity, no value, that is not at its mercy." These fears intensified after Jay
Gould acquired control of Western Union in 1881. Company officials countered
these charges by claiming that Western Union provided its quotations and financial news to all subscribers equally, favoring nobody. Instead of commerce being at Western Union's mercy, Orton maintained that G&S protected the public
"against the schemes of private and irresponsible speculators." He claimed that
G&S operated its financial news services more as a public trust "than as a source

of profit." In fact, G&S consistently refused to enter into any agreement giving an individual or group exclusive rights to quotations in a particular locality, including boards of trade and chambers of commerce, despite the possibility of making more money doing so.[20]

Although Western Union dominated the ticker market, several competitors sprang up, using antimonopoly rhetoric to gain customers and access to the floors of stock and commodity exchanges. Not surprisingly given its status as the nation's leading exchange, NYSE was the major battleground between G&S and its competitors. G&S enjoyed a monopoly over NYSE quotations until 1873, when the Manhattan Quotation Company arose to challenge it. The opposition company promised to reduce rates for ticker customers and to pay the NYSE an annual fee, making it popular with both NYSE members and officers. The two companies engaged in a rate war, dropping the price for ticker service from twenty-five to ten dollars a month, a rate that nearly tripled the number of customers taking NYSE quotations (from both companies) from about four hundred to nearly eleven hundred by the end of 1873. However, this lower rate did not pay expenses and it drove the Manhattan to the wall. By the end of 1875, G&S had driven it from the field and raised rates back to twenty-five dollars a month. For its part, NYSE officers seized the opportunity that the Manhattan's competition had offered. In May 1873 the exchange demanded an access fee of three thousand dollars plus a royalty of fifty cents per ticker per month from each company. G&S objected strenuously that it had never before paid rent and that the Manhattan was an unproved upstart, but it had no choice but to accede. In 1877 the Atlantic and Pacific Telegraph Company, a Western Union competitor controlled by Jay Gould, threatened to start up a competing ticker service. Again, NYSE seized the opportunity to demand more money from G&S, eighteen thousand dollars a year. A final competitor, the Commercial Telegram Company, appeared in 1884 and had ties to the Postal Telegraph Company, the most successful and longest-lived Western Union rival. In 1890 NYSE assumed control of that company, renaming it the New York Quotation Company and making it the quotation provider for exchange members.[21]

While NYSE officers believed that the exchange deserved a financial stake in the quotations generated on its own floor, its demands for increasing rent and royalties seemed like rank opportunism to Western Union's managers. By 1902 G&S had paid the exchange a total of some $560,000 in rent and royalties, and that year the exchange increased its access fee to $100,000 a year. This was a substantial share of G&S's revenue. That year the company's income from quotations from *all* stock and commodity exchanges was about $1 million.[22]

Thus, from 1873 onward, relations between Western Union and NYSE became adversarial, and NYSE's behavior made the telegraph company far less willing to cooperate with the exchange's efforts to deny its quotations to bucket shops. In the absence of meaningful cooperation from Western Union, NYSE decided to control the flow of quotations from their origination on the exchange floor to their final printing on members' ticker tapes. In 1885 NYSE replaced the reporters for both G&S and Commercial Telegram with its own reporters, who gathered the quotations from the floor before turning them over to the telegraph companies. NYSE thereafter required that all those who sought its ticker quotations had to get permission from exchange officials rather than the telegraph companies. NYSE's takeover of Commercial Telegram in 1890 gave it an independent ticker network that supplied its members directly. From this point onward, and in contradistinction to the Chicago Board of Trade (CBOT), the exchange grew far less interested in fighting the bucket shops except for those in New York City that drew business away from its members.

A final class of ticker service was horse racing and sporting events. In an 1887 letter to the postmaster general, Western Union president Norvin Green claimed that the telegraph was the servant of commerce—illegitimate commerce at that. He specified three classes of customers that heavily used the ticker: speculators in commodity options and futures contracts, bucket shops where patrons wagered on the price movements of stocks and commodities, and pool rooms that took bets on horse races and other sporting events. None of these activities, he concluded, "can scarcely be legitimate commerce."[23] At first glance, it is surprising that Western Union's president frequently admitted that his company knowingly abetted gambling. Green intended his statements to reduce the force of the postal telegraph movement. His argument was that the telegraph, as demonstrated by its customer base and message traffic, was a specialized communications medium serving a small market and should therefore be left alone. It is striking that a majority of Western Union's ticker business came from gambling in bucket shops, horse-race gambling parlors, and saloons—and that its president openly acknowledged this in congressional testimony.

Telegraphic coverage of sporting events—and its connection to gambling—dates at least to an 1849 boxing match between Thomas Hyer and James Sullivan. Early professional baseball was also prone to gambling. At the start of the 1876 season, the National League's Chicago White Stockings insisted that Western Union cease supplying local pool rooms and saloons with the results of its home games. Western Union complied but still made out-of-town results available to gamblers. One Chicago gambling syndicate supplied a dozen saloons and pool rooms with three updates during games. Explained one pool room owner, "I propose to know in advance how a game is going if I can, and then I will fleece the fools who come to my place to bet." In 1886 Western Union supplied ticker service to an Atlanta pool room called "The Base Ball Exchange." This pool room specialized in games of the Southern Base Ball League and posted play-by-play results from the ticker onto a blackboard. A senior G&S official explained a few years later that "the gin-mills subscribe very liberally to this class of service." Indeed, the surviving records of the Western Union offices at Chicago and Louisville showed that company managers were well aware that many of their major customers were running gambling dens.[24]

It is difficult to say exactly how much bucket shop business Western Union did, as its accounts did not distinguish between bucket shops and legitimate brokerages. But the company kept separate statistics of its income from its sporting ticker service, almost all of which came from pool rooms and saloons. In his 1890 testimony to Congress, Green claimed that Western Union took in $700,000 a year from New York City pool rooms alone. Between 1893 and 1903, Western Union's total income from horse racing and sporting events results was just under $8 million out of a total ticker income of just over $17 million. Thus, nearly half of G&S's ticker income came from sports gambling. An internal AT&T memo in 1909 thought that "formerly the race track and pool room business was the most profitable service handled by the W.U. Company" and that it amounted to some $2 million a year, including revenue from circuits leased to gambling syndicates.[25]

In 1904, under pressure from Jay Gould's daughter Helen (a major stockholder) and antigambling activists, Western Union agreed to cease supplying pool rooms directly with sporting news. However, the company continued to profit from gambling and thus assisted the rise of modern organized crime in the early twentieth century. After the company ostensibly got out of the pool room business, a former Western Union operator, John Payne, leased a network of circuits to transmit racing news to pool rooms and saloons. In 1911 Chicago crime boss Mont Tennes, widely known for bombing rivals' homes and businesses, took over Payne's operation. Tennes held a lucrative monopoly over racing news across the country, taking in nearly $8,000 a week from New York and Chicago pool rooms alone. As late as 1950, Federal Communications Commission chairman Wayne Coy told Senator Estes Kefauver's subcommittee on organized crime that Western Union was still getting $1 million a year income from wire leases devoted to racetrack gambling.[26]

The ticker broadcast a steady stream of stock and commodity quotations to subscribers around the country. But a modern financial communications sys-

TABLE 4.3.
Western Union's Income from Ticker Service (in thousands of dollars)

Year	Market Quotations	Horse Racing	Baseball, Boxing, Football	Total
1893	588	729	40	1,360
1894	623	463	45	1,133
1895	699	426	55	1,181
1896	774	515	52	1,342
1897	873	647	80	1,600
1898	887	509	51	1,448
1899	892	646	74	1,612
1900	888	670	70	1,628
1901	928	695	43	1,666
1902	1,027	906	50	1,983
1903	1,104	1,034	72	2,210
1904	1,238	498	71	1,807
1905	1,314		85	1,400
1906	1,210		94	1,391
1907	985		106	1,356
1908	1,090		165	1,256

SOURCE: Statistical Notebooks, WUTC.
Note: "Miscellaneous" category not included. The company stopped supplying horse-racing results directly to subscribers in 1904.

tem required a second network, brokers' private wire leases that allowed traders to place orders in real-time response to ticker quotations.[27] Western Union was unwilling to lease circuits to private parties until 1879 because it did not have the spare bandwidth and because its managers were at first concerned that private wires would cut into its ticker quotation business. The company reversed this policy because the quadruplex gave it excess circuit capacity and because brokers' wire leases in fact generated greater demand for tickers. Brokers, particularly the heaviest users of the telegraph, began to set up their own private wire networks to connect their branch offices.

Western Union's first wire lease was in July 1879 to the brokerage house of Edward K. Willard & Company for a line to Saratoga to serve its wealthy customers who summered there. During the 1880s, brokers and hotel proprietors at Newport, Saratoga, Coney Island, Long Branch, and other summer resorts eagerly began leasing seasonal wires. Demand for private wires was so immediate and intense that by the end of 1879 Western Union president Norvin Green complained that he was "pressed every day for new private line contracts and Exchange wires." The company charged about $50 per mile per year or about $15 per mile for seasonal circuits to summer resorts, with stockbrokers accounting for nearly all the year-round wire leases. By 1883 Western Union leased about thirty private circuits to brokers. For example, Henry Clews's brokerage had a Broad Street headquarters, six other Manhattan offices, and branches in Chicago, Philadelphia, Washington, Baltimore, Richmond, Boston, Providence, New Bedford, and Fall River, all connected by private wires. At the turn of the century, Leland & Ware, a large commission brokerage, paid Western Union $200,000 a year for its leased wire network. At that time 136 NYSE members had offices outside of New York connected by private wires. Green and the company's managers remained skeptical of wire leases for several years, noting, for example, in the 1885 annual report that leases accounted for only 5 percent of revenue and that there was "little profit" in that business. Two years later, however, the company's annual report alluded to the "large increase in leased wires to the press associations and private parties." By the turn of the century, one-eighth of Western Union's traffic passed over leased circuits and that sector was a major source of the company's growth in income. By 1915, 80 percent of Western Union's wire lease mileage and income came from bankers and brokers, with most of the remainder coming from press associations.[28]

Effect of the Ticker on Financial Markets

The telegraph network and ticker dramatically improved intermarket coordination and trading efficiency. Thanks to the ticker, the time required to place long-distance trades dropped to a few minutes, even for transatlantic trades through the Atlantic cables. As a result, price differentials all but disappeared in securities and commodity markets. (Price differences remained in spot commodity markets because of differing regional crop supplies and the time and cost of transporting produce between market centers.) As geography increasingly became irrelevant in financial markets, the large Chicago and New York exchanges came to dominate their respective markets. As early as the 1880s, it was increasingly evident that smaller regional exchanges were becoming secondary markets, merely responding to the stream of CBOT and NYSE quotations reaching their floors. Some of this market consolidation occurred by design. In the 1890s, for example, NYSE's officers banned its members from trading its listed securities on other exchanges. By 1910 more than 90 percent of all bond trades and two-thirds of all stock trades in the country occurred on NYSE's floor.[29]

As sweeping as these large-scale changes were, the ticker also reshaped the geography and psychology of financial markets for traders. Indeed, by 1880 the ticker had become so essential to the operation of stock and commodities markets that it had become nearly impossible to envision that trading had been done any other way. In 1903 Sereno S. Pratt, editor of the *Wall Street Journal*, conceded that there had been "active speculation" before the ticker, "but it is difficult to conceive now of a market deprived of its use and compelled to rely upon quotations carried by brokers from office to office. The very life of the Street seems to depend upon accurate, immediate, and continuous quotations from the Stock Exchange."[30]

The ticker offered market participants and financial institutions several capabilities that the telegraph network alone could not.[31] Most obviously, the ticker provided subscribers a continuous flow of quotations, a far more useful and efficient way of distributing market information than prior methods. Equally significant was the profound psychological hold of the ticker on market participants. Some of the ticker's allure arose because it provided constant, real-time financial information. In this sense, its appeal was similar to the increasing demand for frequent news bulletins discussed in the previous chapter. The ticker also mediated between speculators and the mysterious, often inscrutable, workings of the market, a market that could suddenly enrich or impoverish participants. For many traders and speculators, the market was no longer the floor of the Stock Exchange—it had become nothing more or less than the ticker tape. Indeed, almost all contemporary discussion of the ticker emphasized its powerful attraction. Horace L. Hotchkiss recalled that when the ticker first entered commercial service in December 1867, it "created a sensation as the quotations made their appearance on the tape. The crowd around it was at least six deep." Daniel Drew's career showed this psychological hold quite clearly. In November 1869 Jay Gould and James Fisk attempted a corner in Erie Rail Road stock, and Drew, as was his wont, took a short or bear position in the stock, losing some $2 million to Gould and Fisk. Twenty years later, Wall Street broker Henry Clews recalled that Drew, a devout Christian, took spiritual counsel from a friend who advised him to pray. "He tried hard to pray," recounted Clews, "but his irresistible desire to keep constant watch on the tape of the ticker" prevented him. As Clews described the scene, Drew exclaimed to his spiritual counselor, "It is no use, brother, the market still goes up." Even after he retired to upstate New York in 1876, the aged Drew still returned to New York City and was a frequent guest at the Hoffman House, "where he could still have ready access to the ticker." In 1879, at the age of eighty-two, he returned to New York permanently. His son rented him a house on East 42nd Street and leased a ticker for him so that he could speculate with modest sums. For financial writer George Rutledge Gibson, writing in 1889, the careers of ruined speculators like Drew, the so-called ghosts of Wall Street, served as allegories for the ticker's addictive allure: "The ticker is always a treacherous servant. In the end it proved itself the master. Now the man who once dealt in thousands of shares of stock sits in a dingy, little bucket shop," still intently watching the ticker. In 1908 a writer in the financial magazine *The Ticker* diagnosed a common mental disorder caused by the instrument: "speculitis." *The Ticker* published "A Song of the Ticker" in 1910.[32]

Oh, the ticker, oh, the ticker,
How it makes me long to dicker
With the fascinating tape.

Oh, the ticker, oh, the ticker,
How I yearn to get rich quicker,
And from penury escape.

Oh, the ticker, oh, the ticker,
How it sometimes makes me sicker
Than a dope-bewildered ape.

Oh, the ticker, oh, the ticker,
How it makes my wise friends snicker
When they see me wear the crape.

Oh, the ticker, oh, the ticker,
How my senses reel and flicker
When it hits me in the nape.

Oh, the ticker, oh, the ticker,
When it makes the coin come thicker
How I glow and gasp and gape.

Oh, the ticker, oh, the ticker,
How my jaded nerves get slicker
When I'm winner on the tape.

Another reason for the ticker's appeal was that it liberated market participants from being physically present on the exchange floor. This spatial decoupling had several consequences. To begin with, it widened the circle of participation in financial markets and made brokers' offices sites of intense activity. Consider the experiences of Edward Neufville Tailer, a wealthy New York dry-goods merchant who kept an extensive diary and scrapbook for most of his life. Tailer first began speculating in stocks during a rising market in the spring of 1879, and he typically took long positions, hoping that stocks would rise in value. He rarely if ever sold short. He traded through two brokers, Louis T. Hoyt and William Breese, both of whom were his friends. When in New York Tailer monitored the market and placed trades in person at both offices. At times he recorded his regret at not being present to witness large market swings. During his extensive travels on business and vacation, however, he kept track of his investments mainly through quotations published in newspapers. His diary does not indicate that he monitored the ticker while out of town. Indeed, during panics and bull markets he often cut his trips short to return to New York to watch the ticker in person. In

June 1882, for instance, Tailer noted that a 5 percent drop in the market prevented him from attending a friend's funeral. In November 1882 he cut short a business trip to Chicago because of another decline in the stock market. He spent the day after his return at Hoyt's office, watching the panic unfold over the tape.[33]

Tailer's experiences revealed how ticker service reflected the geographic structure of the nation's financial markets in the late nineteenth century. While the major Chicago and New York markets were well on their way to becoming national in scope, exchanges still retained their regional character. For example, in a famous 1888 corner on the Chicago Board of Trade, wheat prices rose more than a dollar a bushel within a few days, but prices on the New York Produce Exchange rose only a few cents. Although Western Union's network was capable of supplying New York and Chicago quotations nationally, ticker subscriptions reflected this regionalism. In 1886 only 160 out of 479 Western Union tickers in the Midwest handled New York stock quotations; of these 160, 109 were in Chicago. As late as 1907, only 90 of the 1,103 tickers in Western Union's New York District handled Chicago grain quotations.[34]

Thus, serious traders like Tailer needed to be present at brokers' offices to monitor the market and react quickly to its changes. This tied Tailer to Hoyt's and Breese's offices during active markets. Tailer was unable or unwilling to keep track of his investments while in Chicago or Newport. After private wire leases became more common, however, brokers opened branch offices in major cities and resort areas, allowing active traders to place trades in response to changing market conditions. For example, the famed bear speculator Jesse Livermore was on vacation in Atlantic City when the 1906 San Francisco earthquake hit. He immediately stepped into the local office of his New York broker, E. F. Hutton, and sold short four thousand shares of Union Pacific. Yet even Livermore thought it difficult to speculate successfully outside of New York, noting that it was impossible to ensure "how close the ticker's prices are to the actual prices on the floor of the Exchange. By the time a man gets the quotation here and he hands in an order and it's telegraphed to New York, some valuable time has gone."[35]

The ticker also gave traders an anonymity that had not existed before and permitted the spectacular corners and manipulations characteristic of the late nineteenth century. Daniel Drew's career shows how the ticker made market manipulations easier. Drew had begun speculating in 1844 and by the Civil War he had become one of the most successful traders on Wall Street. In 1862, as the wartime wave of speculation in stocks and gold gathered strength, Drew set up a small brokerage house, Groesbeck & Company, to serve as his personal broker. Drew and his brokerage were the first adopters of the ticker in late 1867, and he first used it extensively during the so-called Erie War of 1868 that pitted him and his allies Jay Gould and James Fisk against Cornelius Vanderbilt for control of the Erie Rail Road. Clifford Browder, Drew's biographer, described the tactics used by both men. Vanderbilt ensconced himself in an office on West Fourth St. "where early each morning, hidden from the press and public, he issued orders to an army of underlings, . . . who in turn commanded clerks, brokers, secretaries, and runners." Drew, in contrast, occupied "a back room at Groesbeck's" where the ticker "let him follow the market with ease."[36] Although Drew's victory over Vanderbilt was mainly due to his issuance of fraudulent stock and legal maneuvering, his pioneering use of the ticker fit his predilection for secret stock manipulations.

The ticker made it easier for speculators to attempt corners, particularly on the Chicago Board of Trade, the country's largest commodity exchange. Corners became a permanent fixture on the Chicago Board of Trade after the Civil War and were the stuff of exchange legend for the next fifty years, until federal regulation of grain exchanges arose during World War I. Arising from three features of modern commodities trading—futures contracts, standardized grading of grain, and the continuous stream of ticker quotations—they revealed at once the glamour and the greed in speculation. Social reformer Henry Demarest Lloyd exposed the mechanics of grain corners to the readers of the August 1883 issue of the *North American Review*. "Corners," he wrote, "used to come on the Board of Trade once in a year or two. Now there are corners almost all the time. The Chicago corner used to be the venture of some local Titan and was felt only within the then provincial jurisdiction of the Board. Now it is often the cosmopolitan work of the combined capitalists of half a dozen cities, and its effects . . . are felt in advancing prices all over the world."[37]

The Kershaw wheat deal of 1887, one of Chicago's most famous attempted corners and subsequent crashes, showed how market manipulators were able to use the ticker both to maintain their anonymity and to conduct their operations at a distance. The corner had been in surreptitious progress for more than a month when traders realized in mid May that an unknown clique was steadily driving the price of June wheat upward. Trading through the commission house of C. J. Kershaw and Company, they controlled the market completely throughout May and into June. According to Charles Taylor, the board's official historian, "rumor was rife as to the identity of the wheat manipulators." A few members thought that John Mackay, the Nevada silver king, was responsible. Others guessed that it was Phillip Armour, the Chicago meatpacker and architect of several previous corners, "because he was out of town and it was said that he always left the city when he had any large manipulation in hand." Some traders guessed, correctly as it turned out, that a group of Cincinnati speculators stood behind the deal.

The corner broke on 14 June as wheat flooded into Chicago's grain elevators and the price sank dramatically. Only afterward did it become known that E. L. Harper, vice-president of the Fidelity National Bank of Cincinnati, had masterminded the failed corner and had conducted it through the telegraph and the ticker. In the aftermath, nineteen Chicago grain brokerages, including Kershaw, failed. Panics ensued on the New York, St. Louis, and Milwaukee exchanges, and depositors ran on Harper's bank, which failed a week later. Harper himself received a ten-year prison sentence for embezzling funds from the bank to conduct his speculations.[38]

The apotheosis of the continuing abstraction of financial markets was the rise of so-called technical trading or chart analysis, a stock trading method that became popular at the turn of the twentieth century. This type of trading relies extensively, if not exclusively, on charting trends in the quotations of particular stocks, with little or no reference to general economic or market conditions. The technical trader assumes that these general conditions are already reflected in the stock's price. For example, in 1898 Lewis C. Van Riper, an early advocate of chart analysis, urged traders, "Drop all sentiment, pay no attention to new gossip, points, or tips, but merely become a machine with sufficient power to execute your orders according to market movements." The most famous example of chart trading is the so-called Dow Theory, set out in numerous *Wall Street Journal* articles by Charles H. Dow just after the turn of the century. As Dow's disciple William Peter Hamilton explained some twenty years later, "The thermometer records actual temperature at the moment, just as the stock ticker records actual prices." Describing "tape reading" as a "sort of sixth sense," Hamilton urged readers to think of a long time-series of ticker quotations as a "barometer" that predicts future stock prices. Jesse Livermore was one of the most successful technical traders using long time-series as well as up-to-the-minute quotations to conduct his trades.[39]

The Rise of Bucket Shops

The ticker laid the foundation for the modern brokerage industry, solidified the financial power of the major New York and Chicago exchanges, and reconfigured the geographic and psychological relationship between markets and participants. Perhaps the ticker's most significant and lasting effect, however, was to popularize speculation through the institution of the bucket shop. Bucket shops were betting parlors in which patrons wagered small sums on the price movements of stocks and commodities. Bucket shop proprietors typically took the other side of customers' trades and thus did not place these transactions on any stock and commodity exchange. Although bucket shops served as a kind of shadow market for stock and commodity speculation, bucket shop wagers did not affect actual prices of stock shares or agricultural products. Bucket shops first arose in New York in 1877, and within a few years they had spread to Chicago, Milwaukee, St. Louis, and other commercial centers. Within a decade, nearly all states (West Virginia being the leading exception) declared bucket shop transactions to be illegal gambling wagers.

Besides the spreading ticker network, bucket shops owed their origin to the difficulty ordinary people faced in trying to invest through regular brokers. In 1870 financial writer James K. Medberry summed up the scrutiny faced by prospective stock traders: "Brokers are very careful of new men. No respectable firm will take a commission from any one who has not been properly introduced. They want to know all about his means, . . . in fact, everything bearing upon his business peculiarities." Women and people of color, of course, were all but barred from brokerages. Thus, at first, as the Board of Trade's official historian Charles H. Taylor noted in 1917, bucket shops "were not viewed with particular alarm," but as "a sort of democratized Board of Trade, where the common people could speculate." Indeed, they catered to customers, as one newspaper reporter put it in 1887, that "no broker would care to have."[40]

Although exchange leaders were at first unconcerned, bucket shops soon came to draw a great deal of business away from legitimate brokers, particularly their most speculative clients who traded on margin. Because bucket shops operated at the legal and moral periphery of the economy, closed or changed names and locations frequently, and left behind no business records, it is difficult to determine their exact extent. Yet evidence—gleaned from the archival records of Western Union, the Chicago Board of Trade, and the New York Stock Exchange; state and federal court cases; newspaper and magazine stories; memoirs of speculators and financial writers; and exposés by antigambling crusaders and muckraking journalists—shows that the shops were widespread and were the main point of entry for ordinary Americans to speculate.

Although these sources mainly describe how bucket shops operated and seldom gave specific details about their customers, it is possible to draw a composite picture of bucket shop patrons and to trace how bucket shops popularized speculation. Many contemporary accounts emphasized the diverse clientele of the bucket shops. For example, the *Chicago Tribune* in September 1879 described the bucket shop as a place where "no broker is necessary, any person, man or woman, boy or girl, white, black, yellow or bronze can deal directly." However, the first bucket shops sprang up in the financial centers of major cities and catered to the young male clerks who worked in nearby banks and investment houses. These men lacked the means to trade in stocks and commodities, yet they had ready access to financial information and moved in a heady speculative atmosphere. In 1879 the *National Police Gazette* described the bucket shop clientele as composed of "all classes of men who have been bit by the scorpion speculation," but it asserted that the majority of patrons were "young clerks" employed by bankers and brokers who had thus "become imbued with the spirit of stock-gambling." Before the arrival of the bucket shops, "there was no other outlet for this spirit." Jesse Livermore, for example, made his first thousand dollars trading in bucket shops before the age of sixteen while working as an office boy at the Boston brokerage Paine Webber. Five years later, the magazine complained that bucket shops had become "the cause of no end of petty thefts on the part of office-boys and small salaried clerks" to cover bucket shop bets. Newspapers frequently carried stories of respectable professional and family men who lost their savings or embezzled from their employers to cover their losses.[41]

By the mid-1880s, bucket shops had moved into neighborhoods away from financial districts, into smaller cities, and into the countryside. Several uptown New York bucket shops catered exclusively to wealthy women, who preferred to deal there instead of going to Wall Street, where they feared "exciting adverse comment." In 1884 the *New York Times* reported that bucket shops were "thriving in all of the large towns and cities from New-York to Chicago, and that operators who once traded legitimately with members of the Stock Exchange, now operate" through them.[42]

These early bucket shops catered to those who were already predisposed to speculate but could not do so through brokers because of limited means, sex discrimination, or other reasons. By the turn of the century, however, bucket shops not only had spread geographically but were aggressively competing with brokers through newspaper advertisements, printed investors' guides, and tip sheets. As the bucket shop industry expanded into new regions and tapped new customers, it exhibited a trend toward consolidation and concentration. As early as 1887 the *New York Times* reported that a syndicate popularly known as the "Big Four" controlled the bucket shops in Manhattan, had branches in all the large towns of the country, and possessed millions of dollars in working capital. By the turn of the century, the Haight and Freese Company, specializing in New York stocks, operated some seventy branches on the Eastern Seaboard, extending as far south as Richmond and as far west as Pittsburgh and Buffalo. The Coe Commission Company of Minneapolis, specializing in grain quotations, operated one hundred offices across the northern states, from Boston to Spokane. The M. J. Sage Company of New York, specializing in cotton quotations, controlled two hundred branches in the south. The larger offices of these chains made annual profits of $100,000 to $500,000.[43]

While most bucket shop business came from those who wished to speculate with a few dollars, by the turn of the century many new accounts belonged to novice investors. For example, one of Haight and Freese's customers, Ridgway Bowker, a sixty-year-old Philadelphia typesetter who had never before invested in the stock market, became the subject of a muckraking exposé of bucket shop operations. In the summer of 1903 Bowker read a newspaper advertisement for the firm and decided to investigate further. After reading its slick *Guide to Investors*, Bowker opened an account with $150. After some initial successes, he soon incurred losses and found himself handing over more and more money "to protect his margin." He ultimately lost $3,200, most of his retirement nest egg, and was forced to return to work at a salary of $60 a month. Another customer, Charles Weiss of Boston, lost more than $5,000 believing he was investing in stocks; his widow Anna sued in Federal court in 1905 to recover his losses. Merrill A. Teague, the author of a series of magazine articles featuring Bowker, claimed that these incidents were hardly unique. Teague estimated that bucket shops annually took $100 million from "Americans of small means" like Bowker and Weiss.[44]

Bucket shop proprietors were able to lure people like Bowker and Weiss because they claimed that their operations were identical to those of regular brokers, except that they catered to the small investor. To emphasize these similarities, bucket shop proprietors outfitted their offices with the same tickers, blackboards, telephones, and reading matter found in brokers' offices. Jesse Livermore recalled that the Cosmopolitan Stock Brokerage Company, a large Boston bucket shop, had "a fine office and the largest and completest quotation board I have ever seen anywhere. It ran along the whole length of the big room and every imaginable thing was quoted . . . everything that was bought and sold in New York, Chicago, Boston, and Liverpool." In 1886 the *New York World* reported that many of Manhattan's two hundred bucket shops were fitted up as sumptuously as Wall Street brokers' offices. Bucket shop patrons showed "no more diffidence going into these places than is shown going into the New York Stock Exchange, and the class of customers has greatly improved."[45]

A typical bucket shop transaction had the outward form of a margin trade placed with a broker but with much lower margins and lot sizes. Exchanges like the New York Stock Exchange required minimum margins of 10 percent and minimum trades of one hundred shares, transactions involving hundreds or thousands of dollars. By contrast, most bucket shop transactions involved sums from ten to fifty dollars. For example, the Haight and Freese chain charged a commission of an eighth of a percent and required minimum margins of 3 percent per share of stock, three cents per bushel of grain, or a dollar per five-hundred-pound bale of cotton. The Christie-Street Commission Company, a large midwestern chain specializing in Chicago commodities, charged commissions of an eighth cent per bushel of grain and a quarter cent per share of stock and required margins of one cent per bushel of grain or a minimum transaction of ten dollars. A customer placing a typical buy order at Haight and Freese would "buy" five hundred dollars worth of stock, ten shares of stock at the market price of fifty dollars a share. Assuming a 3 percent margin per share meant that the customer paid fifteen dollars plus commission for this transaction.

Bucket shops made their money by engaging in three related practices. While they claimed to cater to speculators of modest means by requiring low margins, this also meant that they could collect customers' wagers with only small declines in a stock's price. A speculator placing a margin trade on the New York Stock Exchange lost the margin if the stock declined by 10 percent. However, a patron of Haight and Freese lost the $15.00 margin on a stock purchase of $500.00 if the stock dropped only 3 percent, to a share price of $48.50. At this point the patron either could deposit more money to "protect the margin" and keep the transaction open or could simply walk away having lost the $15.00 wager. Furthermore, bucket shops often rigged the game by placing "wash sales" to wipe out customers' margins. Bucket shop patrons tended to be "bulls" and to bet that stocks would rise. When bucket shop proprietors realized that many customers were "long" on a certain stock, they placed orders on legitimate exchanges to sell a minimum lot of the stock at a price sufficiently below its current quotation to "wash down" the price. When the low quotation came through on the ticker, the bucket shop could close out its customers' margins. Jesse Livermore recalled that newspapers referred to a sudden, transitory drop in a stock's price as a "bucket shop drive." In addition to "wash sales," some bucket shops delayed the posting of quotations in the customer room by several minutes and refused orders when they knew that customers would win.[46]

As this description of bucket shop practices suggests, the major difference between legitimate brokers and bucket shops lay in the relationship between customer and dealer. A broker acted as a customer's fiduciary agent to place trades on the floors of the exchanges and to render an accurate accounting of trades. Bucket shop proprietors and their customers were adversaries in a rigged, zero-sum game. Bucket shops derived their income from customers' losses, not from commission fees for placing trades. The bucket shops of Boston, New York, Hoboken, and St. Louis collectively banned Jesse Livermore because of his trading successes, something no legitimate broker would have done.[47]

In contrast to orders placed with legitimate brokers, bucket shop wagers violated state antigambling laws. During the 1880s most states banned stock and commodity transactions in which delivery was not contemplated. This distinction was a source of much contention. Brokers on the regular exchanges claimed that delivery was contemplated in all their trades, yet critics of organized speculation routinely pointed out that delivery rarely took place and that most transactions were settled on the basis of price differences. Bucket shops found a loophole in this provision by requiring customers to sign trading slips stating that they agreed to take or make delivery of the stocks or commodities in which they traded. This loophole also allowed bucket shops to obtain injunctions from state courts to prevent telegraph companies and exchanges from removing their tickers. Because their customers signed slips stating that delivery was contemplated, judges concluded that bucket shops were just as entitled to receive quotations as brokers holding exchange seats.[48]

A final difference between brokers and bucket shops lay in the effect of their transactions on markets. Orders placed with brokers set the prices of stocks and commodities traded on exchange floors. The quotations generated by these orders and transmitted over the ticker network provided the raw material for bucket shop transactions. However, this information flow was unidirectional: bucket shop transactions had no real effect on market prices of stocks and commodities. The "wash sales" by bucket shops had only a transitory effect on market prices, and this limited intervention into the regular markets became more difficult after the 1890s when exchanges tightened their regulations and expelled or suspended members who traded for bucket shops.

Despite these differences between bucket shop transactions and brokers' trades, bucket shops functioned as a shadow or alternative marketplace and drew significant business from legitimate brokers. Because bucket shops did not leave behind business records, it is impossible to obtain precise figures on the scale of their operations. Contemporary sources, however, give a sense of their magnitude. In 1884 the *New York Times* claimed that bucket shops were depriving New York Stock Exchange brokers of $1 million a year in lost commission business. Chicago Board of Trade president Abner Wright similarly estimated that bucket shops accounted for 80 percent of the speculative business derived from a general distribution of the board's quotations. In 1889 the *New York Times* estimated that the patrons of the nation's bucket shops wagered the equivalent of 1 million shares a day. By way of comparison, the average daily volume on the New York Stock Exchange in June 1888 was about 140,000 shares. Indeed, in that year competition from bucket shops for commission business had depressed the value of a seat on the New York Stock Exchange from $34,000 to $18,000 and a seat on the Chicago Board of Trade from $2,500 to $800. In 1905 the Haight and Freese chain had between ten thousand and twenty thousand accounts and a daily trading volume of about 70,000 shares. As late as 1913 William C. Van Antwerp of the New York Stock Exchange estimated that one bucket shop in Buffalo alone dealt in 8,000 shares daily, while all the legitimate brokers in that city traded a total of 11,000 shares.[49]

Telegraph service was essential to the operations of the bucket shops. They made strenuous efforts to secure their wire connections by leasing redundant quotation circuits, employing skilled electricians and telegraphers, even tapping brokers' wires and bribing Western Union employees. In 1884 Western Union's attorney told the secretary of the New York Stock Exchange that it was "a difficult matter" to discover how one large bucket shop obtained stock quotations. The managers of the shop "have a competent electrician and sundry linemen in their exclusive employ and maintain a system of mysterious wires from one point to another, by which when there is any interference with their instruments they may obtain quotations from various places." A few months later, the secretary of the Chicago Board of Trade gave his New York Stock Exchange counterpart details of the operations of a large Chicago bucket shop, the Public Grain and Stock Exchange. This bucket shop chain maintained ten leased wires from its Chicago office to cities in the Midwest, obtained quotations from both Western Union and Baltimore and Ohio Telegraph Companies, and employed ten telegraph operators. In 1887 the vice-president of the New York Stock Exchange discovered that this chain obtained its quotations several minutes in advance of Western Union's regular ticker service, employed twenty telegraph operators, and had expanded to 105 branch offices. An Exchange member claimed that this bucket shop chain rented more than one hundred private-wire circuits at a cost of $200,000 a year. When it failed in 1890, the firm's headquarters employed seventy-five clerks and telegraph operators and had 120 branch offices.[50]

Exchange officials frequently complained to Western Union about bucket shops obtaining their quotations. Yet Western Union's managers insisted that the company had no knowledge of bucket shop operations and had no control over how its ticker customers used quotations. However, records of Western Union, the New York Stock Exchange, and Chicago Board of Trade all confirm that the telegraph company knowingly and eagerly supplied bucket shops with ample wire plant. For example, the record book from Western Union's Louisville office shows that local employees were intimately familiar with the proprietors' names, locations, and business details of the bucket shops they supplied. The record book also shows that after about 1880 the quadruplex gave the Western Union circuit managers a great deal of flexibility in connecting bucket shops in Louisville to the various networks of leased wires controlled by the large chains. The local Louisville bucket shops switched affiliations, closed, changed names, and moved frequently, but the quadruplex gave Western Union managers the ability to perform rapid circuit cutovers and to maintain service with minimal interruptions. Bucket shop wire leases persisted until at least 1910 in Louisville. The office apparently stopped recording them after a federal grand jury indicted Western Union for violating a new law prohibiting bucket shop operations in the District of Columbia. Although Western Union temporarily cut off the bucket shops after this indictment, its reformation was short-lived. In 1913 the New York Stock Exchange appointed a special committee to investigate bucket shop operations, "with special reference" to Western Union's role in supplying them quotations.[51]

In addition to Western Union's connivance, widespread public confusion about the difference between speculation and gambling allowed the bucket shops to flourish. Farmers who blamed eastern financiers for their economic woes saw little difference between speculation in futures contracts on the Chicago Board of Trade and outright gambling. Referring to grain speculators, one farmers' newspaper in 1883 charged, "Their business is gambling, too, and they operate upon the same telegraphy reports that the bucket shops do. . . . The principle of gambling is the same in both places and demands the same condemnation and the same treatment by authorities. . . . Close up the dens." Speculators themselves admitted that there was little to distinguish their activities from gambling. In 1888 Charles Hutchinson, the president of the Chicago Board of Trade, gave an address in which he claimed that speculation on the board benefited both producer and consumer. His father, Benjamin Hutchinson, one of the board's senior members, reportedly reacted to his son's speech by exclaiming to a group of traders on the floor: "Did you hear what Charlie said? Charlie said we're philanthropists! Why bless my buttons, we're gamblers! You're a gambler! You're a gambler! and I'm a gambler!" In a similar vein, the *New York Times* reported on an imminent police crackdown to arrest bucket shop proprietors on the charge of "just plain ordinary gambling." But, the reporter wondered, "suppose some zealous citizen should come along and want to press the same sort of charge against the Stock Exchange?" In 1909 David W. James, a Georgia cotton planter, did just that. James refused to pay his broker nearly $50,000 in losses incurred on the New York Cotton Exchange because he claimed that these transactions were gambling wagers, since he never intended to take delivery but only wanted "to play the market."[52]

Even economists who studied the nation's financial markets expressed doubts about the distinction between speculation and gambling. In 1896 Henry Crosby Emery of Columbia University admitted that "the gaming instinct" was integral to speculation, although "speculation is not mere gambling. Whether it is better or worse than gambling is a question on which opinions will long differ." Legislators, too, found little that was praiseworthy about speculation. In 1909 New York governor Charles Evans Hughes appointed a committee to investigate abuses in organized speculation, and the committee reported that speculation exhibited "most of the pecuniary and immoral effects of gambling on a large scale." Indeed, the committee concluded, "only a small part of the transactions upon the Exchange is of an investment character; a substantial part may be characterized as virtually gambling."[53]

Bucket shop proprietors exploited the public's uncertainty about the difference between speculation and gambling to present themselves as respectable brokers who catered to the small investor. In 1899 Haight and Freese issued a *Guide to Investors* to solicit new business, and claimed that its facilities were "designed for the benefit of THE MILLION" who lacked the capital and experience to invest with high-priced brokers. Indeed, in 1905 the manager of Haight and Freese's Philadelphia office insisted in court that his firm was "a competitor of the New York Stock Exchange." In 1906 *Everybody's Magazine* gave C. C. Christie, the "Bucket Shop King," an opportunity to reply to Merrill Teague's four-part exposé of the bucket shop industry. Christie defended his operation as an "independent" broker by citing figures showing that only about 1 percent of trades on the Chicago Board of Trade resulted in delivery of actual grain. Traders concluded the other 99 percent of their transactions by settling on the basis of price differences, exactly as bucket shops settled accounts with their customers. Therefore, Christie charged, the Chicago exchange was "the biggest bucket shop on earth." He accused the major exchanges of being grasping monopolists seeking "to crush the independents. It is a case of Greed versus Freedom." Teague, in his brief rejoinder, reiterated that bucket shops were gambling dens. However, he admitted, "I'll have no quarrel with Christie about the Chicago Board of Trade. . . . Speculation upon margin is gambling; marginal gambling now holds sway on ALL legitimate exchanges."[54]

The Exchanges Confront the Bucket Shops

Teague's rejoinder to Christie neatly encapsulated the difficulty that the directors of the organized exchanges faced when they confronted the bucket shops. Between 1878 and 1915, the exchanges sought to stamp them out. Bucket shops competed with brokers, particularly for speculative clients who traded on margin. Many brokers keenly felt this competition and pressed the directors of exchanges to take decisive action. More importantly, as Teague suggested, bucket shops cast doubt on the moral and economic legitimacy of speculation. Many Americans saw scant difference between trading on margin through a broker, wagering on price movements in a bucket shop, or gambling at cards or dice. The same acquisitive drive and addictive thrill lay behind both gambling and speculation.

Exchange officials and allied economists slowly and painstakingly constructed a distinction between speculation and gambling as a key weapon in the war against the bucket shops. As it emerged full-blown after the turn of the century, the basis of this distinction was the concept of delivery. Behind trades on exchange floors lay real value, actual stocks or agricultural products that could change hands. And even the most speculative trades helped to produce something else of value, an orderly market as symbolized by the stream of quotations printed by the ticker. On the other hand, bucket shop transactions were counterfeits; no articles of value lay behind them. Bucket shop proprietors were economic parasites who fed off exchanges' quotations and drained their customers' pocketbooks. Denunciations of bucket shops thus partook of the same rhetoric of value used by gold-standard advocates against silverites and greenbackers, with a large and shrill dose of moral outrage.[55]

As exchange officials constructed this distinction between 1880 and the turn of the century, they also tried more practical and direct means to shut down the bucket shops. At first, during the 1880s and 1890s, they treated the bucket shops as business rivals and sought to eliminate them by blocking their access to stock and commodity quotations. However, bucket shops almost always obtained injunctions preventing Western Union and the exchanges from removing their tickers. Judges granted these injunctions because they saw little difference in the methods employed by the exchanges and bucket shops, and they regarded the exchanges' campaign against the bucket shops as merely an attempt to crush smaller competitors.

The Chicago Board of Trade faced both this antimonopoly sentiment and conflation of speculation and gambling in its protracted litigation to maintain control of its quotations. In a series of rulings in state and federal courts between 1883 and 1903, judges continually upheld the right of bucket shops to obtain the board's quotations. The judges in these cases ruled that little moral and economic difference existed between trades on the exchange floor and transactions in the bucket shops and that bucket shops had as much right to the quotations as brokers. In 1883 Illinois state judge Murray Tuley enjoined Western Union and the Board of Trade from removing the tickers of the Public Grain and Stock Exchange. Tuley concluded that the Board of Trade is not "engaged in a moral reform movement. . . . It is competition—not immorality which the Board of Trade is seeking to put down." He accused the board of seeking to establish a "monopoly in the dealing in and brokerage of grain and other commodities." This sentiment, equating the board's operations with those of the bucket shops, lay behind an important legal setback for the board. In 1889 the Illinois Supreme Court ruled that the Board of Trade had only two unpleasant options: to provide its continuous market to all parties wishing them—including bucket shops—or to cease distributing its quotations altogether.[56]

Frustrated at the state level, the board appealed to the federal courts, at first with little success. In one 1903 case, a federal judge ruled that speculation on the board's floor "tends only to excite the gambling propensities of the public. Such is not a species of property which appeals to a court of conscience for protection." A second federal judge held that the board's trades were akin to wagers and, as a result, its activities were "so infected with illegality as to preclude resort to a court of equity for its protection." In a third 1903 case, a panel of three appellate judges ruled that the vast majority of transactions on the Board of Trade were "in all essentials gambling transactions" and that the board itself violated an Illinois statute banning bucket shops. The judges concluded that "the Board of Trade does not come with clean hands, nor for a lawful purpose, and for these reasons its prayer for aid must be denied." However, the board appealed to the U.S. Supreme Court and succeeded in 1905 in obtaining a landmark ruling that gave it and other exchanges property rights over their quotations and allowed them to cut off bucket shops from receiving them. This decision concluded protracted litigation costing the board about $120,000 and spanning twenty-five years, eleven states, and 248 injunctions.[57]

In his majority opinion Justice Oliver Wendell Holmes affirmed exchanges' efforts to distinguish between speculation and gambling. He outlined a working definition of speculation that distinguished between its socially useful and harmful varieties and concluded that the former depended on restricting popular participation in the nation's financial markets. He agreed that transactions on exchange floors, even the most speculative that did not result in actual delivery, were "serious business contract[s] for a legitimate and useful purpose" and not "mere wagers" as bucket shop proprietors had charged. When undertaken by "competent men," speculation was "the self-adjustment of society to the probable," a "means of avoiding or mitigating catastrophes, equalizing prices and providing for periods of want." However, "incompetent persons bring themselves to ruin by undertaking to speculate in their turn."[58]

In addition to this lengthy and expensive litigation, exchanges faced another serious obstacle. Telegraph companies, especially Western Union, were unwilling to cooperate with exchange efforts to stamp out the bucket shops because they were reluctant to lose this lucrative market. This led to frequent and often-fruitless conflicts over the terms of the contracts by which the telegraph companies obtained quotations from the exchanges. In 1883, for example, Western Union refused to accede to the Chicago Board of Trade's demand that it halt ticker service to bucket shops. The telegraph company's president Norvin Green not only claimed that this demand was "impracticable" for the company to carry out in the face of bucket shop injunctions but also lectured that "it would defeat what is most important to the business of the Board of Trade—an extended and wide spread distribution of the latest quotations. It is this that makes orders."[59]

In 1886 incoming board president Abner Wright favored ceasing the distribution of quotations altogether, because he blamed the bucket shops for an 80 percent decline in the board's commission business over the past three years. The board was not yet ready to embrace such a drastic step, but it did so in 1890 after the adverse 1889 Illinois Supreme Court ruling. The blackout shut down few bucket shops; most used quotations from other exchanges or obtained the Chicago quotations through surreptitious means. Because trading proved difficult without the continuous market, many board members lost clients to other exchanges. Those exchanges, in turn, proved unable to coordinate their markets with Chicago's. Opposition to the blackout mounted from within and without. In July 1892 the board reversed itself, and Western Union and the Postal resumed transmission of the Chicago ticker, a clear victory for the telegraph companies.[60]

The New York Stock Exchange similarly vacillated between cutting off its quotations entirely and pressuring Western Union to cease supplying bucket shops. In 1897 the governors of the exchange resolved not to renew Western Union's contract when it expired at the end of June, claiming that the telegraph company had failed to live up to its contractual obligations to cut off the bucket shops. The governors accused Western Union of broadcasting its quotations "to practically every bucket shop in the United States" and concluded that "the aid of the Western Union Telegraph Company is essential to this organized system of fraud, and that aid has not been withheld." The Stock Exchange explored several alternatives over the next few months, including setting up a preferred or rapid quotation circuit for its own members and sending out quotations to nonmembers at infrequent fifteen- or thirty-minute intervals, but in the end its directors decided that they had no choice but to keep providing Western Union with its continuous quotations. One despondent governor told a reporter that the inability of the exchange to find a workable alternative resulted "in a victory for the Western Union. . . . As the ticker matter stands now, the Western Union can practically dictate to us any contract it may desire."[61]

Only the threat of an independent telegraph system owned and operated by the major exchanges brought the telegraph companies to heel. In 1900 relations between the Chicago Board of Trade and the companies again reached an impasse. Unable to secure their assistance in cutting off the bucket shops, the board canceled its 1892 joint contract with the Western Union and Postal Telegraph companies. Resolved not to repeat the indecision of the late 1880s and early 1890s, the board decided to control its quotations at every point, from the telegraph instruments near the pits to the tickers in brokers' offices. At first Western Union's managers were unconcerned; the Board of Trade and the New York Produce Exchange had both threatened to set up independent exchange telegraph networks in 1883 and 1890. This time, however, the board quite seriously planned to connect some two dozen exchanges over these wires by 1905 and was well on its way to doing so when Western Union and Postal backed down in the spring of 1901. The two companies signed a new contract agreeing to the exchange's policy for distributing quotations.[62]

After the mid-1890s, faced with a hostile legal environment and poor cooperation from the telegraph companies, exchange officials broadened their campaign against the bucket shops along three fronts. They first undertook a concerted program of publicity to distinguish their operations from the bucket shops. The Chicago Board of Trade, for example, established a Committee on Promotions and helped establish a Council of Grain Exchanges to carry out this task.[63] At the same time the board embarked on a campaign of internal reform to eliminate abuses in its own ranks. In 1895 it launched a thorough investigation of its members' private wire connections, focusing on twenty brokers with large networks. This investigation resulted in the expulsion of one member and a three-year suspension of another for their telegraphic links to bucket shops. In 1900 the board expelled another five members (including two vice-presidents) and suspended twelve for having connections to bucket shops. Finally, exchanges aggressively investigated bucket shop operations and turned over voluminous evidence to local, state, and federal law enforcement officials to spur them into action. In 1896 the Chicago Board of Trade launched a series of investigations that helped a Cook County grand jury obtain 281 indictments. The board also exhaustively investigated bucket shop operations outside of Chicago, often in cooperation with other commodity exchanges. The board hired several private investigators to examine the wire connections of member firms, to pose as customers at bucket shops, and to get insider information about bucket shop operations from their telegraph operators and clerks. By the beginning of 1899, these investigations had resulted in the closure of 188 bucket shops. From 1906 to 1915 the board sent at least seven investigators throughout the country to gather evidence on bucket shop operations. These investigations were instrumental in exposing and shutting down bucket shops throughout the Midwest and West. Investigators in 1909 even rooted out a Pittsburgh ring of nine Western Union employees—including three managers—who had surreptitiously supplied quotations to bucket shops for five years.[64]

The exchanges' publicity and law enforcement campaigns culminated in a 1909 federal law banning bucket shops in the District of Columbia. Armed with this authority, as well as evidence provided by stock and commodity exchanges, U.S. attorney general George Wickersham appointed a special agent, Bruce Bielaski, to investigate and indict the major bucket shop chains. Using methods pioneered by the Board of Trade's investigators, Bielaski spent ten weeks in early 1910 investigating bucket shop operations in seven cities. His efforts shut down several large bucket shop chains with offices in Washington, D.C., and in his annual report for 1910 Wickersham proudly reported that "substantially every bucket shop in the country has been put out of business as a result of this crusade." At the conclusion of the final court case arising out of this crusade in 1913, the bucket shop was on the road to extinction. By the end of 1915, William C. Van Antwerp of the New York Stock Exchange pronounced the bucket shop dead, thanks to the efforts of the major exchanges and the "support of public opinion, the courts, the legislatures, the public service commissions, and the press." He did not include telegraph companies among the exchanges' allies.[65]

The Ticker and the Rise of the Small Speculator and Investor

Before the ticker and bucket shop popularized speculation in the late 1870s, the public typically viewed the manipulations of professional speculators from the sidelines, as fascinated but disinterested spectators. Stories of the failures of speculators in stocks and gold served as moral warnings about the heedless pursuit of wealth, but the vast majority of Americans had little stake in the operation of the nation's financial markets. After the rise of the bucket shops, however, the speculator of little financial means or experience became a moral and economic problem. As early as 1880 there appeared stories of reckless men who squandered tens of thousands of dollars, bankrupted their employers, ruined their reputations, and destroyed their families.

Contemporaries regarded these incidents as examples of personal moral failure. When C. J. Lawrence asked the New York Stock Exchange to investigate bucket shops with an eye toward prosecuting them in 1883, exchange officials concluded that "it was not incumbent on them" to do so. "The Stock Exchange cannot . . . be held responsible . . . for the acts of swindling dealers in stocks, . . . nor can the heedless dupes of such men look with propriety to the Exchange for redress. . . . they have no stronger claim than the losers at faro, or policy, or any other gambling game." Similarly, a decade later the wife of a Brooklyn man who lost money in a bucket shop sought legal redress from the district attorney, James Ridgway, who replied, "I cannot do anything for you. If your husband doesn't want to lose his money gambling, the best thing he can do is to keep away from such places."⁶⁶

Others, however, blamed margin and futures trading for the prevalence of the gambling instinct in speculation. The exchanges themselves, they claimed, were the root cause of both the bucket shop and the ruin of the small speculator. Reformed gambler John Phillip Quinn offered a typical assessment in 1890. If the commercial exchange were "restricted in its scope to the legitimate purposes of commerce, it is unquestionably of the highest benefit to the business world." However, it was also "a gigantic agency for the promotion of gambling" and the source of the bucket shop evil. Similarly, a federal circuit court judge in 1902, in a ruling upholding the right of the O'Dell Commission Company to continue to receive Chicago Board of Trade quotations, argued that the "bucket shops are the offspring" of the regular exchanges. "When this species of gambling on the commercial and stock exchanges of the country ceases, the bucket shops will disappear, and not before."⁶⁷

Those who believed in the social and economic utility of speculation on the organized exchanges, however, regarded increasing public participation as the main danger facing the nation's organized financial markets and calling them into disrepute. The commissioners appointed by New York governor Charles Evans Hughes in 1909 distinguished, in a way similar to Justice Holmes's 1905 opinion, between "speculation which is carried on by persons of means and experience, and based on an intelligent forecast, and that which is carried on by persons without these regular qualifications." The most effective way to remove the gambling element from organized speculation was to "lessen speculation by persons not qualified to engage in it." To do this, the commission recommended doubling the New York Stock Exchange's minimum margin to 20 percent to discourage speculative trading. In 1911 Columbia University professor Carl Parker recommended the "elimination from the field of speculation of those who are unfitted by nature, financial circumstances, or training to engage in it." Similarly, William C. Van Antwerp of the New York Stock Exchange claimed that the "great evil of speculation" lay with the participation "by uninformed people who cannot afford to lose." And the Council of Grain Exchanges in 1915 pronounced itself "opposed to the assumption of risks by those who are not financially or educationally qualified to speculate."[68]

The opposition of exchange officials to broad public participation explains the popularity of the bucket shops. Barred by high margins, large lot sizes, and hostile brokers and exchange officials, even those who wished to make small stock investments had few opportunities to trade outside of the bucket shops. Sereno Pratt of the *Wall Street Journal* estimated that in 1912 only sixty thousand people placed trades on the New York Stock Exchange, and as late as 1916 only eighty out of the New York Stock Exchange's six hundred brokers accepted trades of less than one hundred shares. Recall that one bucket shop chain, Haight and Freese, claimed to have more than ten thousand accounts in 1902. Some of these accounts, like the ones opened by the retired Philadelphia typesetter Ridgway Bowker and the Bostonian Charles Weiss, belonged to people who apparently believed that they were buying small amounts of stock through a legitimate broker.[69]

Popular participation in the regular securities markets increased dramatically during and after World War I, a result of the widespread purchase of Liberty Bonds and participation in employee stock ownership plans. But the eradication of the bucket shops on the eve of the war was another important factor. The *New York Times* noted a "remarkable increase in the odd lot business" on the Stock Exchange in April 1916. "Thousands" of former bucket shop customers, "practically all of whom were small speculators, have opened accounts with branches of Stock Exchange houses." These new customers were used to the mechanics of trading because of their experience in bucket shops.[70]

The eradication of the bucket shops, the Liberty Loan drives, and employee stock ownership plans all propelled an ongoing shift in popular participation in the nation's financial markets, from speculation in the bucket shops to investment through legitimate brokers. Throughout the remainder of the twentieth century, increasingly dispersed stock ownership became a central feature of modern finance capitalism.

Today more than half of all American households own stock, typically through mutual funds and retirement accounts. A sophisticated communication and information infrastructure has made widespread stock ownership and trading possible. Thanks to the Internet and smartphones, investors can place trades at the touch of a button from almost anywhere. These trades are placed nearly instantly—the NASDAQ Stock Market claims that it completes a trade within one-hundred-millionths of a second of receiving an order. The telegraph network and ticker set these changes in motion by liberating traders from the exchange floor and reshaping investors' expectations and perceptions about participating in financial markets.[71]

CHAPTER FIVE

"Western Union, by Grace of FCC and A.T.&T."

THE TELEGRAPH, THE TELEPHONE, AND THE LOGIC
OF INDUSTRIAL SUCCESSION

In November 1879 the Western Union Telegraph Company and the National Bell Telephone Company signed a contract partitioning the electrical communications industry of the United States. This agreement between the giant telegraph company and the fledgling telephone company was one of the most significant events in American business history. Western Union, possessing much more powerful financial and technological resources, withdrew from the new but rapidly growing telephone market in exchange for Bell's promise not to compete with Western Union's telegraph services. The agreement benefited Bell enormously, as it eliminated a threatening competitor and guaranteed its continued existence and prosperity well into the twentieth century. For Western Union and the telegraph industry generally, the agreement marked the beginning of a long process that would eventually doom telegraphy.

This contract thus marked the starting point of Western Union's decline and of Bell's rapid rise in the long-distance communications market. Because Western Union's failure and Bell's success were both spectacular and significant, they lend themselves, respectively, to narratives of declension and heroism. The standard account runs that Western Union's managers were indifferent to technological innovations like the telephone and were content to stagnate as long as they were able to keep paying modest dividends to the company's shareholders. Bell's success, on the other hand, arose because its managers wisely embraced new technology and possessed prescient strategic vision.

While there is much truth to these standard accounts, the failure of Western Union and the success of Bell must be understood in relation to each other. Despite the 1879 agreement to separate the two businesses, the fates of Western Union and Bell remained closely intertwined over the course of the next century. What tied the failure of Western Union to the success of Bell was a porous and shifting boundary between telegraphy and telephony, between what twentieth-century managers and regulators would later call the record and voice communications industries. Through the contract of 1879, Western Union attempted to keep Bell out of the long-distance communications market, which Western Union's managers viewed as the exclusive province of telegraphy. While Western Union's managers withdrew from the telephone industry for what they thought were sound reasons, they agreed to a poorly drawn contract. In particular, the contract failed to give Western Union an equity stake in local telephone operating companies. More seriously over the long run, the agreement allowed Bell to establish long-distance telephone circuits and to compete for leased-circuit telegraph business.

Western Union thus failed in its attempt to use contractual means to erect a solid barrier between telegraphy and telephony after 1879. Instead, Bell used its market power, geographic reach, and program of technological innovation to penetrate this barrier at key points. After Bell inaugurated its long-distance telephone service in 1886, it actively explored ways to use its widespread plant to compete in the telegraph market. These efforts culminated in 1909, when American Telephone and Telegraph (AT&T) acquired enough of Western Union's stock to give it working control over that company.[1] Over the next few years, AT&T president Theodore Vail and his protégé Newcomb Carlton revitalized Western Union's business, modernized its plant and operations, increased its revenues, and put it on a sound financial footing.

However, AT&T agreed to relinquish control of Western Union under pressure from antitrust regulators in the U.S. Department of Justice. After 1914 federal lawmakers and regulators and Western Union executives attempted to reestablish the divide between the telegraph and telephone industries through regulatory means. These efforts failed because AT&T controlled key telegraph technologies and possessed considerable market power. The Federal Communications Commission (FCC), almost immediately after its establishment in 1934, became an arena in which Western Union and AT&T competed to define the terms of the regulatory divide within the communications industry. For the next thirty years, the FCC was ambivalent about the future of the telegraph industry. Regulators frequently noted that a unified wire communications industry under AT&T's control made sense from the standpoint of technical and economic efficiency, but they nevertheless propped up Western Union because it was the only competitive counterweight to AT&T in the long-distance communication market. AT&T also realized that the presence of Western Union as a weak competitor blunted antimonopoly sentiment. However, competition from Western Union diminished as time passed; by the 1960s, regulators recognized that it offered only token competition.

Western Union and the Early Telephone

During the early and mid-1870s, the Atlantic and Pacific Telegraph Company (A&P), a company controlled by Jay Gould, competed fiercely with Western Union. Although A&P handled only about 15 percent of the nation's telegraph traffic, the renewal of competition in this period led to a burst of innovation in the industry.[2] Western Union's intellectual property strategy, and the business environment within which prospective telegraph inventors worked, helps to explain the company's response to Bell's development and commercialization of the telephone between 1876 and 1879.

Western Union's dominance of the telegraph industry created a demand for breakthrough inventions on the part of rivals and start-up lines. Given the rather high barriers to entry after 1866, aspirant firms relied more heavily than Western Union did on technological innovation. Every telegraph company formed after 1866 touted an invention that it claimed would radically transform the industry. A&P, for example, used a duplex invented by French émigré Georges d'Infreville and an Edison automatic telegraph capable of transmission speeds of one thousand words per minute on short lines. While both apparently gave good service, the company itself was in perpetual financial trouble, and only Jay Gould's 1874 takeover saved it from bankruptcy. The American Rapid Telegraph Company, founded in 1879 by former Associated Press head Daniel Craig, adopted an automatic telegraph invented by George Little. Craig promised that Little's invention would make it possible to telegraph hundred-word letters around the country for a dime each and that eventually his company would charge customers "by the yard" of punched paper tape. The American Rapid went bankrupt in 1884. The most successful of Western Union's rivals, the Postal Telegraph Company, began in 1881 to exploit Elisha Gray's harmonic multiplex system, Chester Snow's compound copper-steel wire, and William A. Leggo's automatic telegraph. By 1886, however, Postal had discovered that Leggo's automatic and Gray's harmonic telegraphs were unreliable in actual service, and the company returned to Morse keys and sounders.[3]

Western Union managers and electricians were much more focused about the pace and direction of telegraphic innovation. The company was remarkably forward-looking in embracing innovations that improved speed, reliability, and message-carrying capacity. William Orton developed a coherent strategy of technological innovation, and in this respect Orton was far more progressive than those who succeeded him. From Orton's death in 1878 until AT&T's assumption of working control in 1909, Western Union's three presidents (Norvin Green, Thomas T. Eckert, and Robert C. Clowry) all regarded telegraphy as a mature business that could little benefit from technical improvements. This turned out to be a self-fulfilling prophecy. Western Union did not adopt or develop any significant innovation between Edison's quadruplex and the multiplex printer developed around 1920.[4]

Yet Orton's outlook on technological innovation was fundamentally conservative as well. During the 1870s, Orton and other Western Union managers remained steadfast supporters of the traditional Morse key-and-sounder method of transmission, and they consistently rejected "rapid" or "automatic" transmission systems as having little or no speed advantage.

Furthermore, they tended to buy telegraph inventions or develop them in-house only when subject to competitive pressure and, in many cases, did so only to block competitors from using them. Western Union managers used the company's market power to negotiate hard with inventors, trying to buy their inventions as cheaply as possible. For example, at the height of competition with A&P in early 1875, Orton offered Kentucky doctor Henry Nicholson $25,000 for his quadruplex design. Nicholson thought the price too low and turned down the offer. Three years later, after Western Union and A&P had formed a noncompeting pool, Orton rejected a renewed offer from Nicholson, telling him that "telegraphic inventions of doubtful utility will not . . . find so ready a market as heretofore during the period of active and earnest rivalry between competing Telegraph Companies." While Orton may have been right in not buying Nicholson's quadruplex or adopting automatic telegraphs, the company's niggardliness almost lost it the rights to Edison's quadruplex and played a role in its loss of the telephone market to the Bell interests.[5]

Western Union sought inventions that improved the message-carrying capacity of its lines, particularly automatic relays that increased the range of message transmission and multiplex systems (like the duplex and quadruplex) that permitted the simultaneous transmission of multiple messages on one line. The history of these multiplex systems helps explain Western Union's general approach to telegraphic innovation and to contextualize its relationship to the early telephone. Western Union's adoption of the quadruplex occurred at roughly the same time as the development of the telephone, and the telephone was itself an outgrowth of work on multiplex telegraphy, particularly harmonic telegraphs capable of handling more than four messages simultaneously. Furthermore, the histories of the quadruplex and telephone show that Western Union was reluctant to adopt new technologies, or at least to pay inventors adequately for them, unless prodded by competitive pressures.

In 1868 Joseph Stearns of the Franklin Telegraph Company, a small company operating lines between Boston and New York, devised the first practical duplex.[6] The Franklin used Stearns's invention sparingly, so in 1872 he sold the rights to Western Union. Western Union's successful use of Stearns's duplex spurred its rivals to find another duplex design that did not infringe the Stearns patent. To forestall this, Orton hired Edison in February 1873 to devise as many multiplex methods as he could. By early April Edison reported back that he had tried twenty-three different designs, ten of which he deemed practical and patentable. In the course of his investigations, Edison devised a "diplex," or method for transmitting two messages simultaneously in the same direction. The quadruplex was the combination of Stearns's duplex and Edison's diplex. Edison's work for Orton succeeded in protecting the key Stearns patent. Edison had covered the ground so thoroughly that Western Union held the rights to all duplex methods except d'Infreville's, and Orton was confident that that design infringed upon Stearns's.[7]

In the years that followed, Western Union's managers frequently stated that Edison's quadruplex was the most important telegraph invention since Morse's. The patent was easily worth more than $10 million to Western Union.[8] Orton knew early that the device was valuable, writing to Joseph Stearns in December 1874 that it "realizes all we promised" and had forestalled Daniel Craig's efforts to start an Automatic Telegraph Company. Yet, Orton continued, he had "not settled definitely the question of compensation" to Edison because he had not yet taken out a patent and the quadruplex's capabilities on very long circuits had not been fully demonstrated.[9]

Orton's foot-dragging almost cost Western Union the quadruplex. As Edison later recounted, he had trouble getting access to Western Union's lines to refine his quadruplex, so in desperation he offered the company's electrician George Prescott a half share in the patent to gain this access. In early October 1874, Edison demonstrated that he could work his quadruplex reliably between Boston and New York and, in December, between Chicago and New York. Yet despite Edison's frequent entreaties, Orton refused to discuss purchase terms with him. In need of funds, Edison offered the quadruplex to A&P, which eagerly accepted it for $30,000 in early January 1875. A few weeks later, Orton belatedly agreed to one of Edison's propositions, for $25,000 down and royalties of $233 per year for each circuit created. (However, Edison never received any payment from

Western Union because of the lengthy litigation that resulted from A&P's and Western Union's conflicting ownership claims.) Fed up with his treatment at the hands of Western Union, Edison joined A&P as its chief electrician, a post he held until the end of the year, although Orton was able to restore a decent working relationship with him by July.[10]

After Edison's quadruplex, a major focus of inventive activity in the telegraph industry was acoustic or harmonic multiplexes. These promised to permit more than four simultaneous messages by using several reeds or tuning forks responding to specific acoustic frequencies. Although these never worked well under real-world operating conditions, several inventors demonstrated in the laboratory or in controlled field tests harmonic telegraphs capable of subdividing a telegraph line into ten or more channels. The telephone arose out of Elisha Gray's and Alexander Graham Bell's work on harmonic telegraphs. It was a short conceptual step for both men from transmitting musical tones to transmitting the human voice, although reducing this concept to practice was a considerable technical challenge. The question of priority, whether Bell or Gray deserves credit for the telephone's invention, has been a matter of controversy since 1876. It is most likely the case that the telephone was a case of simultaneous and independent invention deriving from both men's work in harmonic telegraphy.[11]

Gray was an accomplished telegraph electrician and manufacturer who had close personal and business ties to senior Western Union managers, particularly Anson Stager. Both Gray and Gardiner Hubbard, Bell's financial backer and postal telegraph advocate, regarded harmonic telegraphy, not the telephone, as the lucrative invention that would transform the telegraph industry. Bell, on the other hand, was largely unknown to the fraternity of telegraph electricians. Bell, perhaps having the advantage of being a young outsider, turned out to be far more interested in developing the telephone. In July 1875 he had demonstrated, however crudely, that he could transmit sounds with a prototype telephone. Only Hubbard's prodding (and threats to withhold his daughter Mabel's hand) kept Bell working on harmonic telegraphy.[12]

In February 1875 Hubbard arranged a demonstration of Bell's harmonic telegraph for William Orton, who was impressed enough to invite Bell to New York for another round of tests at Western Union headquarters. Bell eagerly accepted. He wrote his parents in early March that A&P had just bought a quadruplex patent for $750,000, allowing it to "compete successfully" with Western Union. (Bell was mistaken—A&P had just paid Edison $30,000 for uncertain rights to the quadruplex.) Because his system, he claimed, was capable of creating thirty to forty channels on a single telegraph wire, he anticipated great results from his demonstration. At first his visit to Western Union headquarters came off brilliantly. Company electricians George Prescott and Gerritt Smith assisted Bell with his demonstrations, and Orton seemed favorably impressed. Later in the day, however, Orton's tone changed. He began talking very favorably about Gray's harmonic telegraph, claiming that Bell's was crude alongside it. Still, he impressed upon Bell Western Union's power to "make the weaker party the stronger," implying that if Bell's terms were favorable, the company would adopt his over Gray's. Orton also deflated Bell's expansive visions of financial reward, claiming that no invention was worth more than $100,000. Finally, he bluntly told Bell that Western Union would never buy Bell's harmonic telegraph as long as Hubbard, the corporation's nemesis on the postal telegraph issue, stood to gain from the purchase. Bell rushed to Hubbard, who offered to stand aside, but Bell would not hear of it. Hubbard then recommended that they take Bell's apparatus to A&P headquarters, directly across the street from Western Union's headquarters, in the morning. Bell went to Western Union to retrieve his instruments and had a final conversation with Orton, who told him that if he took his invention to A&P, Western Union would almost certainly adopt Gray's harmonic telegraph.[13]

Bell came away from his first encounter with Western Union with a healthy dose of suspicion and thought that Orton and Prescott had strung him along. He later claimed that "they had tried to get as much information out of me as possible—and then at the last moment turned me off." Indeed, Bell discovered that in his absence Prescott and Smith had experimented with his device without his knowledge. Furthermore, Orton was a hard negotiator and used Western Union's market power to deflate Bell's expectations of the novelty and value of his invention. A few months later, Orton hired Edison to work on harmonic telegraphs so as to devise methods that did not rely on Gray's or Bell's work. By hiring Edison, Orton obviously wanted to control as much of the harmonic telegraph field as possible in case his company chose to adopt such a system or to block a rival from using the technology. But he also likely wanted to improve the company's negotiating position if it decided to acquire Gray's or Bell's systems.[14]

The Causes and Consequences of the Contract of 1879

Although Western Union's reluctance to pursue harmonic or automatic telegraphs at this time turned out to be sound, historians have long regarded its involvement with the telephone between 1876 and 1879 as a classic example of corporate shortsightedness. The standard account runs as follows. Alexander Graham Bell patented and demonstrated his telephone in the winter of 1876.

Sometime within the next year, in late 1876 or early 1877, Western Union's president, William Orton, reputedly turned down an offer to buy the patent for the telephone for $100,000 from Bell's father-in-law, Gardiner Hubbard. Then, in late 1877, Western Union belatedly realized the value of the telephone and competed fiercely with Bell for the next two years. Finally, Western Union, for reasons of corporate timidity, an inferior patent position, or fear of a competing telegraph company set up by Jay Gould, unceremoniously left the telephone industry in late 1879, settling for a mere pittance of the market's future worth. While there is a great deal of truth to this narrative, much of it has become encrusted with myth, especially the treatment of two major incidents in this history: Western Union's supposed refusal to buy the telephone in late 1876 or early 1877, and Western Union's exit from the industry in late 1879.

On its face, Orton's purported refusal of Hubbard's offer in late 1876 or early 1877 was a blunder of colossal proportions. Although the telephone did not fit Western Union's core markets of rapid long-distance telegrams and financial market reporting, had the company conceived of its mission more broadly, as a provider of electrical communication generally, it ought to have eagerly embraced the telephone. Indeed, for the paltry sum of $100,000, Western Union could have controlled the future telephone industry. Historians have attributed Orton's refusal to Western Union's timidity and its failure to comprehend the new technology's future potential.[15]

This incident of apparent corporate hubris has become so legendary as to make a measured analysis of it difficult. However, the evidence that Gardiner Hubbard ever made this particular offer to Orton is skimpy. Bell's private correspondence neglects the matter completely. Oblique references to ongoing negotiations with Western Union do appear in Hubbard's papers from this period, but neither Western Union nor Bell corporate records make any mention of such an offer.[16] This legendary offer apparently derived from three sources. During congressional testimony in 1883, Norvin Green recalled that Hubbard "came to see me once, and told me that we could take the whole of the Bell patent and control it for $100,000. We declined to do it, and now it has a market value of nearly $20,000,000." In his 1924 memoirs, Chauncey Depew recalled that Hubbard had offered him a one-sixth share of Bell's telephone patent. Depew asked Orton's opinion of Bell's telephone, who according to Depew replied, "There is nothing in this patent whatever, nor is there anything in the scheme itself, except as a toy. If the device has any value, the Western Union owns a prior patent called the Gray's patent, which makes the Bell device worthless." Finally, Thomas A. Watson recalled in 1913 that Orton had "somewhat contemptuously" turned down Hubbard. Yet Watson also claimed that the negotiations were long enough and favorable enough for him to entertain "visions of a sumptuous office" in the Western Union headquarters, "which I was expecting to occupy as Superintendent of the Telephone Department of the great telegraph company."[17]

Depew's and Watson's recollections point to a more likely version of events, that the Bell interests negotiated with Western Union officials over a long period and that Western Union thought that it had the upper hand in these negotiations. Gardiner Hubbard's correspondence in his personal papers and in the AT&T corporate records suggests that he and Western Union negotiated over the course of about a year, from October 1876 to December 1877. Yet Orton's and Green's official letterbooks make no mention of any such negotiations before February 1878, possibly indicating that Western Union's senior managers did not overly concern themselves with the matter or that they did not keep official records of these negotiations.

If Hubbard had offered the telephone to Western Union for $100,000 sometime in late 1876 or 1877, the reasons attributed to the company's refusal are also problematic. Though historians have attributed Orton's refusal to his characterization of the telephone as a useless "toy" and his personal dislike for Gardiner Hubbard,[18] neither explanation suffices. While Western Union may have been slow to come to terms with outside inventors, the company had been investigating the telephone's possibilities since July 1875. At that time, the company hired Edison to research harmonic telegraph methods, including the possibility of transmitting speech. By the following spring it placed Thomas Edison on a $500 monthly retainer to develop a "speaking telegraph." If Hubbard had indeed asked for $100,000 as the selling price of Bell's patent, Orton would probably have regarded this sum as exorbitant for an untried device from an unknown inventor. Orton probably believed that Western Union could develop or purchase a better telephone or play Bell off against other inventors to secure a lower price for his patent.[19] Historians have also claimed that Orton disliked Hubbard because he was at the forefront of the postal telegraph movement. Indeed, Orton and Green spent a great deal of time fighting off various nationalization schemes. At one point in the 1870s, Orton and Hubbard appeared so frequently before congressional committees that one wag dubbed Congress the "Wm. Orton and Gardiner Hubbard Debating Society."[20] Although the two frequently clashed in public on this issue, their personal relations were respectful and cordial. However, Hubbard's advocacy of a government telegraph had irritated other Western Union officials, who were determined that Hubbard should not gain financially from any prospective deal.[21]

From a short-term perspective, Western Union's rejection made sense—the telephone did not fit with Western Union's core markets. The telephone could not make a written record and private-line telegraphs already served the needs of Western Union's customers requiring short-distance two-way communication. And because the telephone could not be multiplexed, its widespread adoption would have reduced the company's message-carrying capacity, particularly in these high-traffic urban areas. Western Union's managers were not alone in their early failure to recognize the telephone's possibilities. Elisha Gray, himself a telephone inventor, commented in late 1876 that he and other "practical telegraph men" regarded Bell's telephone as "a beautiful thing in a scientific point of view," but when seen "in a business light it is of no importance." Because "speed" was paramount, the harmonic telegraph held far more promise than the telephone. Yet, Gray conceded, "if perfected," the telephone might "have a certain value as a speaking tube." Similarly, British Post Office telegraph engineers Henry Fischer and William H. Preece used Bell's telephone during an extended fact-finding mission to the United States in the spring and summer of 1877. They found that they could converse through it, but "the result was not so satisfactory as experiment led us to anticipate." They thought that the telephone would improve with time, but in its present state it was of no "practical value," and they did not recommend that the British Post Office adopt it. In September 1877 Gardiner Hubbard reported that several entrepreneurs seeking to set up local telephone companies tried Bell's telephone and found that they "cannot use them." They required temperamental adjustment to overcome interference from wires on crowded urban telegraph poles and therefore did "not work very well." The public remained skeptical as well. Bell spent most of 1876 and early 1877 demonstrating his telephone to lecture hall audiences, and his company did not lease telephones to paying customers until April 1877.[22]

Western Union's managers likely wanted the Bell interests to assume the risk of being first movers and to enter the market when the telephone had demonstrated some commercial viability. During the summer and fall of 1877, several of Western Union's local managers, most notably Anson Stager in Chicago and Henry Bentley in Philadelphia, began to set up Bell exchanges. By August the Bell company had received orders for more than a thousand telephones. To their dismay, executives at Western Union headquarters discovered that many of these early telephone subscribers were replacing their private line printers with Bell's new devices. This spurred Western Union into activity. In July the company commissioned the electrical manufacturer Joseph T. Murray to build four sets of telephones for testing, and in August hired electrician Franklin L. Pope to survey the state of the telephone art. Orton met with Hubbard irregularly while he backed telephone work by Edison, Gray, and Amos Dolbear. By October Gardiner Hubbard had apparently decided against selling Bell's patents outright to Western Union. Instead, he thought that he was close to a deal with Western Union subsidiary Gold and Stock Telegraph Company to lease it telephones on favorable terms. The deal fell through, according to Hubbard, because Orton "want[ed] the whole" of Bell's patent rights.[23]

Soon afterward, Orton and senior Western Union managers decided to compete with Bell actively. Their decision stemmed from their inability to reach agreement with Hubbard for the sale of Bell's patents and from Pope's and Edison's conviction that Gray's and Edison's work on telephony predated Bell's. In December Western Union organized the American Speaking Telephone Company (AST), two-thirds of which was owned by the Western Union subsidiary Gold and Stock and one-third by the Harmonic Telegraph Company (founded in 1876 to exploit Gray's harmonic patents). In January 1878 AST marketed its first telephones. AST's organizers offered a quarter share of the company to Bell. The Bell interests countered with a half share of the consolidated company plus a dollar per year per telephone royalty. Negotiations broke down over this impasse in February 1878 and did not resume for more than a year.[24]

Over the next two years, after the telephone had proved its worth, Western Union competed fiercely with Bell through its AST subsidiary. Western Union established exchanges, recruited subscribers, and developed a better telephone than Bell's. The Bell interests possessed less capital and resources than Western Union. However, it had a better corps of sales agents in the field, because Bell paid them a 25 percent commission, whereas AST paid only 10 percent. Its major asset was Alexander Graham Bell's fundamental patent claiming the use of an undulating current to transmit the human voice. At the time, however, the legal weight and technical utility of Bell's patents were uncertain. Bell sued Western Union representative Peter Dowd for infringing his patents in September 1878, and testimony did not end until July 1879. In the so-called Telephone Cases of 1888, the U.S. Supreme Court narrowly upheld Bell's patent by a one-vote margin. In contrast, Western Union's cluster of telephone patents covered an instrument that worked as well as, or better than, Bell's telephone. The telegraph giant also had the financial muscle to force Bell out of the telephone market through ruinous competition and expensive litigation.[25]

Western Union's major objective in aggressively competing with Bell in 1878 and 1879 was not to drive the smaller company from the field but to force it into a consolidation. Western Union had done much the same thing a few years before when it reached an agreement with its rival in the financial news service market, the Gold and Stock Telegraph Company. Under the terms of an 1871 agreement, Gold and Stock doubled its capital and gave half of the increased sum to Western Union in exchange for its ticker patents and customer base. This arrangement at once amply satisfied Gold and Stock's shareholders and gave Western Union working control over the company. Western Union pursued a similar course with the Bell company, hoping to effect a similar settlement. This did not happen, however, because the companies could not reach agreement on the relative merits and value of their telephone patent portfolios. Western Union's managers claimed publicly, and their private correspondence suggests that they were sincere, that Gray's patents on harmonic telegraphy controlled Bell's telephone patents. Bell's backers were even more adamant that his patents controlled the technology.[26]

After it became clear that neither side had a clear title to the patent rights that controlled the telephone industry, negotiations resumed again in the spring of 1879. Conditions were more favorable for a settlement than they had been a year previously. Both Bell and Western Union had new management. In April 1878 William Orton died suddenly at the age of fifty-one. His death placed the more tractable and less able Norvin Green at the head of Western Union. William Forbes, an excellent corporate leader, replaced Hubbard and Thomas Sanders at the head of the Bell Telephone Company. Furthermore, Green and Forbes did not have the personal history that Orton and Hubbard had had over the postal telegraph issue. Most importantly, both Green and Forbes were disposed to reach a settlement. Green and the Vanderbilt interests that controlled Western Union were alarmed by the renewal of active telegraph competition from Jay Gould's American Union Telegraph Company formed in May 1879. From Forbes's perspective, competition with American Speaking Telephone was a drain on the Bell company's finances.

Negotiations began in April 1879 with Dr. Samuel S. White acting as go-between. White, a Philadelphia manufacturer of dental equipment, was Elisha Gray's main financial backer. White's plan was to form a new company, the National Telephone and Telegraph Company, capitalized at $5 million. Both Bell and American Speaking Telephone were to get 30 percent of the capital, with the remaining 40 percent to be decided according the relative strength of each side's patents. By the end of June, negotiations had reached a dead end over the patent question. The Bell company wanted to determine the relative weight of each side's patents after testimony had been taken in the Dowd patent infringement suit. Western Union's attorneys, on the other hand, worried that doing so would expose the weak points of both side's patents and embolden new entrants into the telephone market.[27]

Western Union did an about-face in September 1879 when it offered to settle with Bell for only a 20 percent stake in the telephone industry. Historians have typically identified three reasons why Western Union accepted a junior position in the telephone market: entrepreneurial blindness to the telephone's value, recognition that Bell's patents were valid and controlling and that Western Union's telephone infringed them, and willingness to sacrifice its stake in the smaller telephone market to maintain its monopoly in telegraphy in the face of a hostile takeover threat by Jay Gould. Another factor might have played a part as well. According to Edison, the real explanation was that some of the major Western Union stockholders also held large amounts of Bell Telephone stock and were anxious to benefit from their investments. Although no documentary proof backs up Edison's suspicions, this sort of dealing was common at the time.[28]

As Western Union's activities during 1877 and 1878 showed, its managers realized that the telephone held promise. Thus, the company did not exit the telephone industry in 1879 because of bureaucratic timidity or shortsightedness. It can be faulted only for regarding the telephone as an adjunct to the much larger and well-established telegraph industry and for failing to predict that the telephone industry would one day supersede telegraphy in long-distance communications.

A somewhat more satisfying explanation is that by the summer of 1879 Western Union's patent experts realized that Bell's patents were valid and controlling. However, this requires some explanation. Western Union's experts also maintained that its telephone patents covered a cluster of improvements that were necessary for the clear production and transmission of sounds. In other words, each side could conduct a successful telephone business only by infringing the other's patents. Western Union's lead patent attorney, George Gifford, a man with thirty years' experience in telegraph patent litigation, admitted that Bell's patents were controlling. However, he countered that Western Union's patent position was "very strong" and "might cover all known forms of telephones." Had both sides vigorously pressed infringement suits, Gifford believed the result would have been a deadlock that would have blocked either company from offering a commercially practicable telephone. Gifford urged Western Union to seek a settlement on the basis of rough parity, and he regarded Western Union's acceptance of a 20 percent share of telephone rentals during the life of the Bell patent as "a small interest" and "much less than I claimed" in meetings with Bell's attorney. Similarly, at the conclusion of negotiations in September 1879, Green told Bell president William H. Forbes that Western Union's coming to terms was not an admission of an inferior patent position: rather, "our people feel strongly, and are legally advised, that we shall have less difficulty in continuing to use our telephones than you will in yours."[29]

The overriding reason for Western Union's capitulation was the threat of a hostile takeover by Jay Gould after August 1879, especially threatening to Gould's rival and Western Union's largest shareholder, William H. Vanderbilt. In May 1879 Gould set up a competing telegraph company, the American Union, capitalized at $10 million, and used his extensive railroad holdings to build a national telegraph network quickly. A strong and determined competitor, the American Union ate into Western Union's revenues and depressed the value of Western Union's stock, which Gould quietly bought. Furthermore, if Western Union wished to regain its monopoly, it had to buy out its rival; doing so gave Gould more Western Union stock. In this way, Gould ultimately succeeded in acquiring control of Western Union in 1881.[30]

While using the American Union as the main front of his drive to control Western Union's stock, Gould also sought an alliance with the Bell interests, to which Bell's managers were quite receptive. Gould probably had several objectives in mind by threatening Western Union with this alliance. He generally sought to reduce Western Union's profits and the value of its stock by competing with it as broadly as possible, and an alliance between Gould and the Bell interests presented a formidable threat to Western Union's managers and stockholders. Also, local telephone exchanges generated substantial telegraph business—an arrangement with Bell would have diverted customers wishing to send telegrams from Western Union to American Union. Indeed, managers of both Bell and the American Union realized the benefits of an alliance against Western Union, and they actively explored such an arrangement in the spring and summer of 1879. At the same time, Gould obtained control of several telephone exchanges in New England and had begun connecting them to each other over the wires of the American Union.[31]

The value of this potential alliance for Bell became clear in September 1879, when Green and Vanderbilt decided to remove the threat it posed by acceding to Bell's terms. This was a sudden and total shift in Western Union policy regarding Bell and the telephone. As recently as July, Norvin Green had directed George Gifford, in his ongoing negotiations with Bell's attorneys, to "insist . . . on the larger share of the joint Company," or at the least "an equal partnership." In mid-August, Green was still confident that Bell would lose its infringement suit against Western Union agent Peter Dowd and that his company possessed "the only form of Telephone . . . in general and satisfactory use." Yet, in early September, about two weeks after Gould reorganized his American Union Telegraph Company, Western Union came to terms and agreed to a settlement, asking only for a mere 20 percent royalty on telephone rentals in exchange for its patent rights. The final agreement, signed in November, also gave Western Union minority equity stakes in several large telephone exchanges in New York, Philadelphia, Pittsburgh, Michigan, and California and required Bell to turn over all telegraph business generated by it to Western Union. Such a major retreat was hard for many American Speaking Telephone investors to swallow. Green told Forbes that he had "had much difficulty in bringing the interests purely telephonic to consent to this, as they considered it too small a share in the business," but that Western Union's board consented to this agreement "for the sake of peace and harmony, and to avoid the trouble, loss and expense of a bitter and wasteful competition." Green thus sacrificed Western Union's future in the telephone industry, most likely at Vanderbilt's behest, in order to fend off Gould and to regain its monopoly in telegraphy. Gould obtained control of Western Union anyway in 1881.[32]

Viewed strictly as a financial investment, the settlement did not reflect the true value of Western Union's telephone property and patents, but it did generate substantial revenue for the company. In addition to receiving a $1 annual royalty per telephone until 1896, Western Union came away with substantial stock holdings in the industry, worth more than $2.5 million in 1880. During the mid-1880s Western Union earned an annual income of about $400,000 from its telephone investments, and its income between 1882 and 1899 totaled about $6.4 million. Western Union also shed much of its equity stake in the telephone industry in order to finance new telegraph construction without using earnings earmarked for dividends. Thus, Green and Gould used the company's holdings in the telephone industry as a cash cow to maintain dividends to shareholders in periods of low profit margins. Western Union's profit margins were low in this period because of competition from upstart telegraph companies, especially the Postal Telegraph Company, and because of prevailing antimonopoly sentiment, which fed a resurgence in the movement to nationalize the telegraph.[33]

However, neither side got quite what it wanted from this contract. Bell's managers frequently complained that Western Union used its holdings in the major telephone exchanges to delay and obstruct Bell's expansion. Almost immediately, in early 1880, Bell officials encountered "unusual inaction" on the part of their Gold and Stock counterparts to implement the terms of the contract. While Green seemed disposed to carry out the contract in good faith, major Gold and Stock shareholders remained convinced that it had been a bad bargain.

Over the next few years, Western Union withheld money due Bell, particularly dividends on Western Electric stock and commissions on telegrams sent or delivered over the telephone. In 1883 Western Union brought suit against Bell's California operating company to prod Bell into buying its remaining telephone properties for $2.1 million, a figure that Forbes found exorbitant. In 1885 an exasperated Forbes accused Green of "looking for technical grounds for a basis of litigation" to revise the terms of the 1879 contract.[34]

Forbes's charge had merit to it. Western Union soon came to realize that it had overlooked two glaring flaws in the contract: the agreement failed to give it a 20 percent interest in the stock of local telephone companies, and it allowed Bell to operate both a long-distance telephone business and a leased-circuit Morse telegraph business. Thus, the contract, Western Union's first attempt to use legal means to prevent Bell's encroachment on its markets, failed to confine Bell to short-range voice communications, as Western Union had intended.

In January 1883 Western Union resolved to correct the first flaw. At the behest of the directors of the American Speaking Telephone Company, Western Union pressed Bell to turn over 20 percent of the stock of Bell operating companies. Forbes refused on the advice of counsel that there was "no foundation" in the contract language for such a claim. Indeed, Gold and Stock's annual report for 1880 told its shareholders that the agreement gave Bell "exclusive control of the Telephone business" in exchange for "20% of its gross receipts for rentals." Its 1881 annual report noted that the contract had been a good deal, as it resulted in $36,000 in quarterly royalties from telephone rentals with no expenses incurred. Forbes attributed the claim to Jay Gould's machinations, suspecting that he now had designs on the telephone industry and was "casting about for some plausible excuse for a fight, with possible intentions upon the price of the stock." But Western Union was in fact serious, and in 1883 it sued Bell for 20 percent of the shares of Bell's licensees. Because of the accounting complexities involved and the vagueness of the contract on this point, the suit dragged on for thirty years. The U.S. Supreme Court in 1913 finally decided in favor of Western Union, but it was a hollow victory. The Court directed AT&T to pay Western Union a mere $5.3 million, the amount of the original value of the stock in 1879 plus interest and dividends. The settlement did not take into account appreciation of the shares. Because AT&T controlled Western Union in 1913, the telegraph company accepted a settlement for only $1.7 million in compensation. Ironically, the suit cost both sides more than $10 million in legal fees.[35]

While Western Union's failure to get a 20 percent equity stake in the telephone industry was a major financial oversight, its inability to block Bell's entry into the long-distance and leased Morse telegraph business had far more serious long-term consequences. Western Union officials were at first unconcerned; they regarded the telegraph as the better technology because it produced a written record. In 1880 Bell received several inquiries from newspapers about getting news bulletins over the telephone, a clear encroachment on one of Western Union's major markets. Green readily gave his permission, telling Forbes that the telephone could never supply the news as cheaply as the telegraph. More seriously, the contract of 1879 allowed Bell to establish "extra-territorial" lines in order to connect local exchanges but did not permit it to use them to conduct a public message service. However, the contract allowed Bell to lease private circuits over these lines, and within a decade Bell would compete aggressively with Western Union for the private-wire market.[36]

Bell's Encroachment on the Telegraph Market after 1879

Western Union's retreat from the telephone industry in 1879 marked the beginning of three decades of technological and organizational stagnation from which the company never fully recovered. Between 1880 and 1909, Green and his two successors, Thomas Eckert and Robert Clowry, pursued a conservative business policy to secure three limited objectives: to protect Western Union's status as the dominant firm in the telegraph industry, to stave off nationalization of the industry; and to pay a steady, modest dividend. The company expended little effort in stimulating demand for telegraph service or in innovating to improve telegraph technology. As a result, gross income increased only at about the rate that the economy grew overall, but net earnings remained roughly static.

Western Union's managerial and technical laxity placed it in a poor position in the communication market. While it stagnated, the Postal Telegraph Company launched an aggressive campaign to wrest business away from Western Union after John Mackay's son Clarence succeeded him as president in 1902. Of graver long-term concern, Bell rapidly expanded into the long-distance market after 1886 and undertook a far-reaching research-and-development program in long-distance telephone technology. At the same time, it sought to breach the contractual boundary between the telegraph and telephone markets at key points. The complacency of Western Union's managers after 1879, coupled with Bell's aggressive expansion into long-distance communications, put the telegraph industry on a long road to obsolescence.[37]

In 1886 Bell established its first long-distance line between New York and Philadelphia. Almost immediately, it used its long-distance lines to lease private wires extending between New York and Philadelphia for both telephone and Morse telegraph service. In early 1887 Bell's president Forbes offered to give Western Union a 10 percent commission on all public message business it transacted over long-distance lines if Western Union agreed to lift the ban on Bell's conducting such a business. Green, of course, refused. A year later, Green protested to new Bell president Howard Stockton that Bell's customers leasing long-distance lines were sending telegraphic market quotations, often taken directly off Western Union's ticker. When Stockton asked Green to provide specific examples, Green replied that "all the wires leased by Brokers" between Boston and Philadelphia were used for this purpose, and he again reminded Stockton that this was an abuse the 1879 contract "was especially designed to prevent." After three years of fruitless protest, an exasperated Green brought suit, protesting one last time to Bell's executives in early 1891 to restrain its local operating companies from running a public message service. Not only had Western Union's "short-line business around Boston . . . been almost entirely supplanted by the telephone [used as] a message service, [but] the most glaring violation of the contract" was the use of leased long-distance wires for financial quotations, "and this violation is more aggravated by the fact that you actually permit telegraph instruments to do the service by telegraph."[38]

Bell's managers pleaded that it was impossible to restrain the local operating companies from violating the 1879 contract, despite their frequent admonitions. However, these incursions into Western Union's markets were deliberate and were part of a broader, long-term plan to control the long-distance communications industry. Prominent Bell executives in the 1880s—including at least Theodore Vail, William Forbes, Howard Stockton, and chief electrician Thomas Lockwood—all favored pursuing a dominant position in both telegraphy and telephony. Indeed, Theodore Vail recalled in 1906 that he had pushed for the "ultimate absorption of the telegraph business" after the expiration of the 1879 contract in 1896, "that we should ask the Western Union for half of its capital stock for the privilege of continuing in business as one of our subordinate companies." Indeed, the official historian of Bell Telephone Laboratories acknowledged in 1975 that the use of wire plant for both telephone and telegraph service had been a leading goal of Bell research and development throughout its history. At the same time Norvin Green was protesting to Stockton in 1888, Stockton and Forbes were weighing the desirability of engulfing the telegraph industry:

> Now as to controlling W.U.: I think the Telephone and Telegraph ought to
> be worked together and that a consolidation of interests would be of benefit
> to both. The advantages are manifest enough: a complete system instead of

a baulky team, and end to needless losses, —a rapid development of such a system of communication as probably no one outside of the leading telephone men has dreamed of. Something very enticing about it I agree.

Still, Forbes continued, the prevailing antimonopoly sentiment among much of the American public made a "direct acquisition" of Western Union risky. He particularly feared "a serious excess of wrath against monopolies attended by practical legislation . . . resulting in either passage of rate bills or Government Telegraph [and] Telephone management and ownership." Thus, Forbes was concerned that much of the antimonopoly sentiment directed at Western Union would adhere to Bell in the event of a direct takeover.[39]

Because outright acquisition of Western Union seemed risky in the political climate of the day, Bell's managers decided to encroach on the most lucrative telegraph markets by adopting promising telegraph innovations and aggressively promoting its leased-wire business. Both strategies rested on Bell's particular strengths, its door-to-door geographic coverage and its development program to improve the range and quality of its long-distance service. From the beginning of its long-distance service, Bell used high-quality circuits, composed of twisted-pair copper cables that reduced interference between circuits and improved the range of telephone transmission. Western Union, on the other hand, still relied on circuits of a single iron wire with a ground return. Bell's leased-wire subscribers easily divided the twisted-pair telephone circuits into two high-quality telegraph circuits.[40]

Thus, the goal of competing with Western Union for leased-wire telegraph service pushed Bell to sustain a program of technological innovation in order to make more efficient use of its installed wire plant. In addition to research-and-development work to extend the range of its long-distance voice service, Bell adopted a "composite" transmission system that could work a voice circuit and two telegraph circuits on one pair of wires. The Belgian inventor François Van Rysselberghe originally patented this system in the 1880s, and the Bell company refined it and began using it for leased circuits between Boston and New York in the spring of 1890. The system proved so successful that some observers in the mid-1890s were predicting that Bell would drive Western Union and Postal from the telegraph field in a matter of a few years. The composite circuit was the technological basis for Bell's aggressive entry into the leased-wire telegraph market as well as its long-standing claim to future regulators that its telegraph business was merely a by-product of its telephone business and not a major market that it deliberately sought.[41]

Bell's choice to compete for the leased-wire business made sense. It was the most lucrative and growing part of the telegraph industry at the turn of the century. Western Union estimated that $13.76 out of every $20.00, the standard per-mile lease rate, was pure profit, a margin of nearly 70 percent. Between 1893 and 1907, Western Union's revenues from leased wires almost tripled, rising from about $1.3 million to $3.5 million. At about the same time, Bell's private-wire telegraph mileage rose from about ten thousand in 1890 to about seven times that in 1905.[42]

At first, Western Union managers were unconcerned about Bell's competition in the long-distance communications market. Norvin Green told one worried stockholder in 1889 that the long-distance telephone actually benefited Western Union by stimulating telegraph traffic at those points and that Bell's telegraph wire leases were "inconsiderable." Yet by 1894 Western Union had lost nearly all its leased-wire business between Chicago brokers and their New York correspondents. In 1892 and 1893 Bell sent canvassers to major commercial centers, including Chicago and Boston, targeting the bankers and brokers who were the heaviest users of Western Union's leased wires. They discovered that Western Union's leased wires were often out of service, sometimes for several hours during peak business hours, and that the telegraph company was indifferent to customers' complaints. One broker told the Bell salesmen that his leased wire to New York rarely worked for more than half a business day. The issue was the use of the quadruplex to make up these leased circuits. The quadruplex required constant attention to adjust the artificial lines to match changing electrical conditions on the real telegraph lines. Failure to do so rendered the quadruplex inoperable, particularly during wet weather. Conversely, one broker's telegraph operator told the canvassers that Bell's line was the best that he had ever worked. Another complaint was that Western Union often took leased circuits out of service without warning to use them for public telegram traffic during interruptions. These Chicago brokers welcomed competition from Bell, but only if the company could provide comprehensive facilities. Western Union threatened to cut off entirely any customers who leased a Bell circuit. For many large investment houses, getting on Western Union's "black list" would have been disastrous. Bell's canvassers concluded that a large market waited to be tapped but that Bell needed to expand its wire plant in order to switch over potential customers completely from Western Union. By 1894 the company had done so.[43]

AT&T's Acquisition and Control of Western Union, 1909–1914

As the experiences of these Chicago brokers suggest, by the turn of the century the telegraph industry had become moribund. Western Union did not market its services or care deeply about providing high-quality facilities. Customers used the telegraph only because it was a business necessity. At the turn of the century the American population sent about one telegram per year per capita but placed about ten telephone calls per capita. Over the next decade the telegraph industry lost more ground. Between 1900 and 1910, growth in the telegraph industry tracked the country's increase in population almost exactly, while growth in the telephone industry occurred at sixteen times the rate of population increase. The major reason for the stagnation of the telegraph industry was that its leaders had done little to improve service or innovate.[44]

The telegraph industry's stagnation encouraged financiers and industry insiders to consider a grand consolidation of all wire interests after the turn of the century. Consonant with the great merger wave that reshaped the American corporate landscape in this period, such a consolidation would have unified Bell, Western Union, and Postal under one management. J. P. Morgan, who wielded more economic influence than anyone of his generation, stood behind these plans. Between 1902 and 1907, control of Bell shifted from the old cohort of Boston financiers who had dominated its affairs since 1880 to a group of New York bankers headed by Morgan. Although the Morgan syndicate never exercised day-to-day control over Bell's operations, it was responsible for bringing Theodore Vail, who had been Bell's general manager from 1878 to 1887, out of retirement in 1907. Vail and Morgan shared the general conviction that telephone competition between Bell and the so-called Independents was wasteful and ultimately ruinous. More significantly from the standpoint of the telegraph industry, both agreed that the telephones and telegraphs ought to be run under a single management.[45]

Two memoranda from New York attorney and financier John Canfield Tomlinson to Vail in 1903 give a sense of how the Morgan interests planned to proceed. Tomlinson's starting point was the telephone company's large and growing debt. Between 1902 and Vail's 1907 ascension, AT&T's obligations soared from $60 million to more than $200 million. Tomlinson thought that "the financial necessities" of AT&T were "so great and will require such comprehensive treatment that . . . the mere magnitude of the transaction" required consideration of a comprehensive wire merger. He also thought that both Western Union and Postal were ripe for acquisition because telegraphy "afford[ed] no field for development, no opportunity for talent and no attraction to capital." Tomlinson assured Vail that he knew through his business connections that both telegraph companies were "pocketed corporations. Both Mr. [Clarence] Mackay and Mr. [George] Gould have social aspirations and are willing to retire from the control of these corporations," especially if AT&T could guarantee them a 25 percent profit on their stock. Because an outright acquisition might run afoul of Justice Department or Interstate Commerce Commission regulators, Tomlinson suggested a thousand-year lease of both telegraph companies at generous rentals (5 percent on Western Union's stock and 8 percent on Postal's). He concluded by promising that Morgan would "be ready to carry [the deal] out" by early fall 1903. In 1903 Morgan succeeded in naming two allies, Thomas Jefferson Coolidge and John I. Waterbury, to Postal's board of four trustees, but matters did not proceed further.[46]

Although this grand consolidation of all wire interests never occurred, Vail continued to mull over how AT&T could best enter the telegraph industry. In 1906 he summarized his views in a letter to Bell president Frederick Fish. Vail favored acquiring Postal instead of Western Union, because he thought that Postal's domestic telegraph system existed mainly to feed Mackay's overseas cable business. For Western Union the reverse was true. Thus, Vail concluded that "any fight over the domestic telegraph business would result in disaster to the net earnings of the 'Western Union' while it is doubtful if it would be particularly noticeable" to Postal. Whether Bell decided to buy control of one of the existing telegraph companies or to start a competing service, Vail was confident "it would have its own way."[47]

Clarence Mackay, son of Postal's founder John Mackay, agreed that the wire industry ought to be consolidated. He thought he was the right man to do it. He sought control of AT&T by buying its stock on the open market, until he held more than 5 percent of its shares by 1907, more than six times the number of shares held by the next largest shareholder. Mackay demanded a seat on AT&T's board of directors, but Vail and the Morgan interests refused because they wanted to control Postal, and not the other way around. This refusal earned AT&T Mackay's everlasting hostility, and he would later prove key to breaking up AT&T's working control of Western Union in 1913.

Because Postal was essentially a sole proprietorship controlled by Mackay, Vail and the Morgan interests were unable to bring Postal under Bell's sway. However, they succeeded in gaining working control of Western Union in 1909. This came about after the New York Telephone Company announced a reorganization plan that entailed issuing additional stock. Western Union owned about a third of New York Telephone's stock, but because of flat earnings it could not raise the necessary cash to participate in the new stock issue. As a result, its directors felt it prudent to divest its New York Telephone holdings entirely. Also, George Gould, Jay Gould's son and the largest Western Union shareholder, had encountered financial difficulties during the panic of 1907 and welcomed the opportunity to sell his interest in Western Union to raise cash. Gould, Western Union's directors, and AT&T's managers worked out a deal in which Western Union sold its New York Telephone shares to AT&T and Gould sold his Western Union shares to AT&T. This gave AT&T a 30 percent stake and working control of the telegraph company. That the telephone company paid 13 percent above market value for Gould's holdings indicates Vail's (and Morgan's) enthusiasm for acquiring control of Western Union. By 1910 the two companies had seven directors in common, including Vail, a partner in Morgan's investment bank, and three Morgan allies.[48]

As the public and private pronouncements of AT&T's senior managers made clear, this purchase was the culmination of a long-standing goal. Vail had long believed that the telegraph and telephone networks were complementary, that they provided different kinds of service but were best operated under a unified management. Although Vail championed continuing development of long-distance telephony, he thought that telegraphy would remain the cheaper and more popular option for long-distance communication well into the foreseeable future. And the telephone was cheaper than the telegraph for short-haul urban communication. However, because they were both wire communication, they could share plant and personnel. By 1913, such economies had resulted in nearly $400,000 in annual savings. Also, telephone collection and delivery of telegrams could cheapen and popularize telegraphy. AT&T managers called upon their Western Union counterparts to make "every possible effort" to encourage the "filing of telegrams by telephone." By the end of 1913 Western Union customers filed about 11 percent of their messages by telephone and accepted delivery of about 14 percent in the same way.[49]

The unification had immediate benefits for both companies. From AT&T's standpoint, it allowed AT&T to gain control of Western Union's telephone stocks. For years Bell managers had accused Western Union of using its large equity stakes in local telephone companies to obstruct the development and expansion of the telephone business. Another objective was to share Western Union's exclusive access rights to boards of trade, racetracks, and baseball parks. Access to these venues allowed Bell to enter the lucrative financial news and sports reporting markets. Western Union also enjoyed exclusive right-of-way contracts with railroads, which barred potential telegraph competitors, including AT&T, from operating telegraph lines on more than 90 percent of the nation's railroad mileage.[50]

Bell's control of Western Union had great benefits for the telegraph company as well, arresting its slide toward managerial chaos and technological obsolescence. The most visible improvement was the rejuvenation of Western Union's paleolithic management methods. In 1910 Vail hired the accounting firm of Price, Waterhouse & Company to go over Western Union's books thoroughly. Their report pointed out "serious defects" in its accounting standards and the "imperfect condition" of its financial records. On the eve of AT&T's takeover in 1909, administration of the company's operations was indeed chaotic. The company's chief clerk, John Calvin Willever, filed important correspondence and contracts in an old wardrobe cabinet, and the head auditor placed financial reports in piles on the floor of his office. One Bell manager later recalled that the company's headquarters "looked like a page from Dickens." By the end of 1910, Vail had put in place modern accounting practices and streamlined company operations. For shareholders, the most obvious sign of this reorganization was the format of the company's annual reports. Clowry's last report in 1909 had the same format and provided the same financial information as the company's first annual reports from the 1870s. The 1910 report was thoroughly revamped.[51]

AT&T management vastly improved Western Union's technical capabilities as well. Bell's spare circuit capacity was ample enough to meet Western Union's demands for increased facilities for many years to come. Furthermore, AT&T's plant gave Western Union more flexibility in making up temporary circuits for major sporting events, political conventions, and the like. Under Vail's leadership Western Union launched a sorely needed modernization program. The only substantial technical improvements Western Union had adopted since the quadruplex and stock ticker of the 1870s were the introduction of electric dynamos in 1893 and the replacement of its iron wires with lower resistance copper a few years later. Telegraphy still relied on traditional Morse operation, a sixty-year-old technology. Thus, one of Vail's first acts upon assuming management of Western Union was to direct Bell's manufacturing and research subsidiary Western Electric to develop a multiplex printing telegraph to replace Morse keys and sounders. By 1913 these printers had quadrupled message-handling capacity on trunk routes, though Morse transmission remained in place for local service.

Perhaps the most important transformation was a thoroughgoing change in management philosophy. Western Union's management philosophy before 1909, if it could be said to have had one, was merely to retain dominance of the telegraph industry and to pay shareholders a steady if unimpressive dividend. One newspaper reporter quipped that Western Union's employees comported themselves as if they were "doing the public a favor" in transmitting their telegrams.[52] Vail, on the other hand, fully embraced the notion that the communications industry was a public utility. He believed that his company was obligated to give the public high-quality and affordable service, to cooperate with regulators, to provide employees with well-paying jobs and comfortable working conditions, and to safeguard the value of its stock for shareholders of modest means. Upon Vail's takeover, he imbued Western Union's managers with his brand of corporate progressivism. He immediately set about increasing the company's wages, raising them by 50 percent by the end of 1913. He also set up pension and sick-leave plans and improved working conditions. Seeking to shed Western Union's long-standing association with bucket shopping and sports gambling, he cut back on wire leases used for immoral purposes.

The keystone of Vail's public-utility philosophy was load leveling. That is, the telegraph industry had traditionally sought to provide its business customers with rapid if expensive communications. From an operational standpoint, this meant that Western Union needed sufficient plant and operating staff to meet peak demand. Because this peak demand coincided with business hours, much of this plant remained idle at night. In Vail's annual report for 1912, he decried "the effect of irregular loads in the telegraph business" and called for new services that would "equalize and steady the day load and greatly augment the night load." In particular, Vail introduced Day and Night Letters that allowed customers to send longer telegrams for deferred delivery at the same rate as regular ten-word telegrams. These new services proved popular, particularly when coupled with telephone collection and delivery. From the company's standpoint, Day and Night Letters made more efficient use of plant. The results of these reforms were impressive; between 1909 and 1913 Western Union's revenues increased by about 75 percent and the company had regained the public confidence it had lost over the past several decades.[53]

Despite the manifest technical and economic advantages of a unified telegraph and telephone service, AT&T relinquished its control of Western Union at the beginning of 1914. Political and regulatory considerations lay behind this decision. From 1909 onward, Clarence Mackay of Postal protested loudly and continually to federal antitrust officials that AT&T's control of Western Union was illegal. Mackay complained in particular that AT&T discriminated against it by giving Western Union preferential rates for leasing circuits and by connecting telephone customers who wished to send telegrams directly to Western Union offices. Mackay's complaints were justified. Internal Bell memos during this period show that managers frequently refused Postal's requests for circuit leases. One AT&T manager explicitly directed a local operating company not to let Postal "anywhere near our physical plant." In 1912 the New York State Public Service Commission agreed with Mackay that Western Union's telephone collection and delivery privileges were discriminatory and ordered it stopped. When Bell did nothing to change its practices, Mackay threatened that the only solution was "a proceeding by the Attorney General of the United States to separate entirely" Bell and Western Union. Federal regulators sided with Mackay and maintained that AT&T would have to offer Postal the same circuit-leasing rates and telephone collection privileges that it gave Western Union. Because doing so would have eliminated two important competitive advantages for Western Union and might have made Postal into a much more formidable rival, AT&T was unwilling to accede to Postal's demands.[54]

Also, Vail had embarked on an aggressive campaign to acquire independent telephone companies. In 1912 the independents had lodged complaints with the Department of Justice, which filed an antitrust suit in July 1913 against AT&T's interconnection and acquisition policies. Recent antitrust decisions breaking up Standard Oil and American Tobacco portended a similar outcome if AT&T persisted in its policies. Finally, the new Wilson administration was actively considering the nationalization of the telephones and telegraphs, placing additional pressure on AT&T to relinquish its holdings. Thus, in 1913 federal regulators sought to reestablish a divide between the telegraph and telephone industries to prevent AT&T from monopolizing all of electrical communication. AT&T acquiesced in order to stave off antitrust proceedings or government control of the communications industry.[55]

While Western Union came away in 1914 with a revitalized organization and a much needed modernization program, AT&T retained important assets that placed Western Union at a long-term competitive disadvantage. Indeed, the disengagement transformed AT&T from a close ally to a formidable competitor. Although AT&T did not compete with Western Union in the public telegram business, it did skim the cream from the lucrative leased-wire market. This process had begun while AT&T still controlled Western Union. Vail attributed the drop in Western Union's leased-wire business to the decision to cease leasing wires to bucket shops and horse racing pool rooms. By 1914 Western Union had canceled about $500,000 in such leases. However, reputable customers were unhappy with the quality of Western Union's service and sought out Bell circuit leases. In 1912 Bell general manager Charles H. Wilson warned Vice-President Nathan Kingsbury that the Associated Press, one of Western Union's largest customers, wanted to transfer its business to Bell. Wilson suggested that Vail and Western Union general manager Newcomb Carlton settle the matter before Western Union lost more business to Bell. After Bell shed its Western Union holdings, it retained a majority of the leased-wire business, reducing Western Union's share from 70 percent in 1909 to less than 30 percent in 1915. Bell also retained its access to Western Union's railroad rights-of-way, which allowed it to expand its telegraph business at will. By 1915, Bell had replaced Postal as Western Union's main competitor.[56]

The more serious long-term problem for Western Union was that it lagged Bell in key technical areas. Between 1910 and 1913, engineers from Bell and Western Union collaborated on the development of printing telegraphs. After 1914 the two companies continued to pool their patents, but the bulk originated with Bell. A 1922 pooling agreement specified 134 issued or pending patents taken out by Bell's manufacturing subsidiary Western Electric, compared to 39 by Western Union. A 1921 memo noted that Western Union did not have any patents "that are of vital interest" to Bell but that Bell shared patents of "considerable value" with Western Union. Furthermore, the pooling agreements specifically confined Western Union's use of printers to public message telegrams and barred their use on leased wires, a business that Bell wished to retain for itself. Similarly, Bell refused to license its carrier-current patents to Western Union. Bell's technical superiority in printers and carrier-current circuits enabled it to introduce its highly successful Teletypewriter Exchange (TWX) service, which made serious inroads into Western Union's market share after 1931.[57]

The Telegraph Industry after 1931

The fifteen years after AT&T's disengagement were very successful for Western Union. Newcomb Carlton, Vail's handpicked successor, oversaw the completion of an ambitious $200 million modernization program, which replaced the company's antiquated Morse instruments with multiplex printers. Revenue tripled from $46 million in 1914 to $140 million in 1929. But the Great Depression was a serious blow to the telegraph industry, one from which it never fully recovered. Combined revenues of Western Union and Postal peaked in 1929 at about $164 million, dropping to $97 million in 1933. By the late 1930s the Postal Telegraph Company, accounting for about 15 percent of the nation's telegraph business, was bankrupt and kept afloat only by $12.5 million in federal loans. Western Union weathered the Depression but saw its absolute and relative position in the long-distance communications market decline precipitously and permanently.

TABLE 5.1.

Relative Positions of Western Union, Postal, and Bell in the Public Telegram and Leased Wire Markets, 1918

Company	Income (thousands of dollars)		
	Leased Wires	Public Telegrams	Total Telegraph
Western Union	1,065 (27%)	38,640 (69%)	39,725 (66%)
Bell	2,219 (57%)	9,411 (17%)	11,630 (19%)
Postal	626 (16%)	7,852 (14%)	8,478 (15%)

SOURCE: From "Private Wire Contracts," case 5421, *Interstate Commerce Commission Reports*, vol. 50 (Washington, D.C.: GPO, 1918), 735.

By World War II, one former Western Union executive lamented, telegraphy had become "a beer and pretzels business." FCC chairman James Lawrence Fly told Congress in 1941 that telegraphy was a "very sick" industry, stopping just short of calling it "dying."[58]

Several factors besides the general business downturn of the 1930s brought the telegraph industry to this state. Perhaps the major reason for its long decline after 1929 was the public's increasing preference for other long-distance communications media, such as the toll telephone, airmail, and Bell's TWX system. All of these offered either better service or lower prices. For instance, a twenty-word telegram and a twenty-word reply between New York and San Francisco cost just over four dollars, but a three-minute station-to-station telephone call cost slightly less. When AT&T disengaged from Western Union in 1914, the telegraph industry handled about 80 percent of the long-distance communications market. In 1931 the telegraph and telephone each handled a roughly equal share of the nation's long-distance communications. In 1940 Western Union and Postal combined handled about 23 percent of intercity communications, with AT&T handling about 73 percent, and the small but growing airmail service accounting for the remaining 4 percent. Even in the telegraph market itself, Western Union's market share was declining. As early as 1921, AT&T accounted for 90 percent of the lucrative leased-wire and teletype markets, with Western Union and Postal splitting the remaining 10 percent.[59]

The telegraph industry might have been able to stem this tide had it developed an internal research-and-development capability to keep pace with the state of the electrical art. In contrast to AT&T's strong support for these activities, Western Union devoted only 0.4 percent of its annual budget to them. Inattention to research and development hamstrung Western Union in the two major classes of telegraph technology: terminal equipment and circuit capacity. As early as 1921, AT&T believed that Western Union's lack of development and manufacturing capabilities for terminal equipment put it "in a decidedly unfavorable position." Western Union president Newcomb Carlton refused to buy out its vendor of teletype equipment, Morkrum-Kleinschmidt (later renamed the Teletype Corp.), for $1 million in the early 1920s because he did not want Western Union to expand into equipment manufacturing. In October 1930 AT&T bought the Teletype Corp. for $30 million. In that year alone, Western Union bought $9 million of equipment from the company.[60]

The situation was similar with respect to circuit capacity. Bell developed carrier-current technology as an outgrowth of its research on electronic amplifiers for long-distance telephony. The new technology allowed one pair of wires to carry five simultaneous telephone conversations or twenty telegraph transmissions. Stated differently, carrier-current channels cost about five dollars per year per mile, while traditional circuits cost about six times as much to operate. Bell had started installing carrier circuits in 1918, and Western Union had recognized their usefulness as early as 1921 but was barred from using the technology until the last Bell patent on electronic amplifiers expired in 1931. Bell refused to license carrier-current technology to Western Union, because, as Bell engineer M. C. Rorty warned, doing so would have given the telegraph company enough extra circuit capacity "to compete on equal, or perhaps even superior terms" for telegraph wire leases.[61]

Bell's monopoly on carrier-current technology arose directly from its research-and-development arm, a capability that Western Union lacked. Bell's carrier-current technology dramatically increased the bandwidth of AT&T's wire plant, making it possible for it to handle all wire communications at little additional cost. As one internal Western Union memo phrased it in 1933, this excess bandwidth was "a powerful urge" for AT&T to "develop new services" like its Teletypewriter Exchange service introduced in 1931. Although Bell refused to share its patents with Western Union, it leased spare carrier channels to the telegraph company after 1928. Although the telegraph company would install a national microwave transmission network after World War II, it relied on AT&T circuit leases for the majority of its capacity until the 1980s.[62]

Thus, by 1930, Western Union was in the awkward position of buying terminal equipment and circuit capacity from its biggest competitor. Its long-overdue modernization, begun by Carlton in 1915, was obsolete when it was completed in 1934, superseded by AT&T's TWX and carrier channels. TWX allowed a subscriber to connect through a central exchange with any other subscriber and to carry on a two-way printed "conversation." TWX, in other words, combined the advantages of the telephone and telegraph by allowing two-way direct communication and by providing a written record. TWX subscribers could hold a three-minute, two-way printed exchange consisting of about 180 words for the same price as a 24-word telegram at full rates between the East and West Coasts. Western Union and Postal immediately countered with a competing timed-wire service at roughly the same price, but it found little favor with business customers because it could not provide two-way service. Bell's introduction of TWX was a source of great consternation to Western Union's managers. Although Bell claimed that the conversational feature of its TWX service made it simply an extension of its telephone service, Western Union's managers and many telegraph customers regarded it as a decisive step in a Bell plan to dominate the telegraph industry. According to an internal Western Union report, when Bell introduced TWX, "the market value of Western Union securities suffered severe declines, several large banking houses made guarded inquiries about the Telegraph Company's future, and a number of estates and investment trusts removed Western Union securities from their portfolios of sound investments." In 1939 former FCC official Noobar Danielian more succinctly called the service a "deathblow to the telegraph companies."[63]

Because of the success of Bell's TWX and carrier-current technology, Western Union finally set up an internal research-and-development capability in the late 1930s. The company's engineering vice president, Frederic d'Humy, prodded President Roy White several times between 1935 and 1938 to launch a $40 million plant modernization and research-and-development initiative over the next decade, focusing on developing a microwave transmission network and a system of facsimile transmission. In particular, d'Humy thought that fax could leapfrog AT&T's TWX and compete with airmail:

> Without reservation I foresee FACSIMILE as a prime mover for transmitting the "written word" by telegraph. This transition is inevitable. It is only by the aid of this means that Western Union will be able to regenerate an ascending place in the field of communication

D'Humy's goal was to be able to offer customers a fax system that could transmit sixty words for the price of a three-cent postage stamp. Doing so would divert a large percentage of airmail onto Western Union's lines. Although White, a former railroad president, had a reputation as an aggressive cost cutter, he directed d'Humy to develop fax technology. By 1940 d'Humy pronounced fax technically perfected and ready for commercial introduction, although its full market rollout had to wait until after the war. By the mid-1950s, Western Union had placed some thirty thousand DeskFax machines in service.[64]

Although d'Humy warned White not to "depend too strongly" on the Federal Communications Commission to roll back AT&T's encroachments, by 1940 federal regulators had decided to prop up the telegraph industry as the only counterweight to Bell's long-distance telephone monopoly. Western Union managers charged that AT&T's "policy of vast expenditure and aggressive competition" in telegraphy showed that it intended to dominate the telegraph market, creating a near-total monopoly in wire communication. Many regulators agreed that the FCC needed to take action to prevent this from happening. In a standard textbook on the economics of regulation published in 1936, one FCC regulator called Bell's competition in the telegraph industry "vicious," because Bell could sell its service "at by-product rates, it being a very small part of the total business of that system." Former FCC official Noobar Danielian charged that the Radio Corporation of America (RCA) and AT&T had formed a "Radio-Bell Axis" to divide up the telegraph industry between them. The implication was that only a complete regulatory separation of the two industries would prevent Bell from using its market power to compete unfairly with the telegraph companies.[65]

Between 1938 and 1943, Congress held five hearings to investigate the precarious state of the domestic telegraph industry. The solution that emerged in these hearings was to urge Western Union to buy out Postal. Representatives of the Department of Justice's Antitrust Division and the FCC agreed that such a merger was the only way to revitalize the telegraph industry, or at least to ensure that it did not fall into the hands of Bell or disappear entirely. However, official FCC policy was ambivalent in its commitment to preserving the telegraph industry, and it wavered between considerations of technical and economic efficiency and the public interest. On the one hand, officials recognized that a single integrated domestic wire monopoly made a great deal of sense and that Bell could dominate the telegraph industry in short order if given a free rein. On the other hand, the FCC recognized that such a move would create serious repercussions for telegraph employees, for the public, and for the nation's defense. In May 1941 FCC chairman Fly summed up the commission's position before Congress. He conceded that there were "serious arguments of an engineering and economic nature" for an AT&T monopoly on all wire communications, but he and the FCC opposed such a monopoly on "broader economic grounds and on social grounds." In particular, the FCC wished to preserve a healthy Western Union to provide some measure of competition to AT&T.[66] This ambivalence would characterize the FCC's policy toward the telegraph industry until the deregulation movement of the 1970s.

Federal regulators and industry observers universally approved the 1943 merger of Western Union and Postal as necessary and long overdue. But Western

Union acquired Postal reluctantly, under heavy pressure from the FCC. The bill permitting the merger required Western Union to sell off its overseas cable operations, to assume Postal's $12.5 million debt to the federal government, and to give Postal's staff generous employment guarantees. Although FCC officials also pressured AT&T to sell its leased-wire and TWX operations to Western Union, AT&T withstood heavy regulatory pressure to retain them until 1971.

The merger with Postal and the upsurge in traffic during World War II brought a temporary respite to the beleaguered telegraph industry. A few years after the war, however, it became increasingly clear that Western Union's survival still hung in the balance. While FCC regulators tried to prop up Western Union as a competitive counterweight to AT&T, they were consistently pessimistic about the telegraph industry's prospects. By 1949 Western Union handled only 16 percent of the country's long-distance communications, and the company was losing $1 million a month. The FCC officially informed Congress that it was doubtful whether "any program which can be devised can save Western Union." In response, Western Union's president, Walter P. Marshall, lobbied Congress and the FCC to make Western Union the "chosen instrument" in what he called the "record communications industry." Marshall, an old telegraph hand with three decades of experience at Postal, sought a federally regulated monopoly over all domestic and international communication involving the production of a written record, including AT&T's TWX service. In contrast, AT&T executives did not wish to part with TWX, and they argued instead that Western Union's proper sphere of operations was restricted to "telegraph service," which they narrowly defined as "accepting, transmitting to a distant point, and delivering a message."[67]

Marshall's intensive lobbying failed to move Congress or the FCC to anoint Western Union as the country's "chosen instrument" in domestic and overseas telegraphic communication. The FCC would not permit Western Union to buy out its international telegraph competitors and reiterated its position that the company needed to sell off its existing overseas operations. Marshall did not succeed in convincing the FCC to remove the regulatory barrier between the domestic and international telegraph markets; however, he did persuade the FCC to accept his more expansive definition of "record communications" to describe his company's business, instead of AT&T's more restrictive label of "telegraph service." From the 1950s onward, federal regulators increasingly referred to Western Union and AT&T as the "record" and "voice" industries, respectively, and they continued to prod AT&T to sell Western Union its TWX service.

Until the end of Marshall's tenure in 1965, Western Union would remain dependent on the FCC and AT&T for its survival. Despite an ambitious modernization program to build a nationwide microwave transmission system, in the early 1960s Western Union still leased 60 percent of its circuit capacity and bought most of its terminal equipment from AT&T. One business writer in 1959 aptly summed up Western Union's predicament in an article entitled, "Western Union, by Grace of FCC and A.T.&T."[68]

A more serious weakness was Western Union's dependence on its public message service for the bulk of its income, a business rapidly becoming obsolete and irrelevant in the communication marketplace. Marshall had some limited success reducing the company's dependence on the declining message telegram market by tapping new markets related to its core competencies but not subject to FCC regulation, particularly customized communications systems it set up for government and industry. In 1950, for instance, Western Union set up a teleprinter network connecting nearly two hundred major banks, and a facsimile system for the U.S. Air Force to connect more than two hundred bases. The specified rate of return on the Air Force contract was 7.5 percent a year, far higher than the return the company could expect from the regulated telegraph service. He hoped to generate $500 million in revenue from these private leased networks by 1969.[69]

Marshall rescued Western Union from the verge of bankruptcy and obsolescence. Although the company still relied on message telegram service for the majority of its income, about two thousand companies were leasing customized communications networks from Western Union, revenues and earnings were at all-time highs, and the company's debt was retired or advantageously refinanced. With its financial house in order and its modernization complete, Marshall planned to spend another $350 million by 1965 in new plant and equipment to transmit both television programs and computer data communications. By the early 1960s, Marshall was betting Western Union's future on data transmission as the growth market of the next decade. In sum, Marshall succeeded in transforming Western Union into a modern telecommunications provider through massive yet responsible spending.[70]

In 1965 Marshall recruited his replacement, Russell McFall, a forty-three-year-old manager from the glamour industries of aerospace and electronics. McFall shared and amplified Marshall's vision for Western Union's future. He planned to remake Western Union into "a national information utility" in which customers would lease time on Western Union mainframe computers, with the terminals and transmission lines all owned by the firm. He also wanted to establish a comprehensive satellite communications system to transmit television programming and other broadband applications. As McFall explained shortly after assuming the presidency, "We are in the communications service business, but this entails a new concept of communications. We have to broaden communications to include the processing of information as well as the handling of it." Yet he admitted that a very real problem was "how to build these systems without spending so much money that we kill ourselves in the process."[71]

Thus, the course that McFall chose for Western Union posed a dilemma. In order to grow as a modern telecommunications company, Western Union needed to compete in the data-processing market. Doing so meant taking on much larger corporations, especially IBM and AT&T, a task requiring a great deal of capital. Because the company still relied on low-margin message telegrams for the majority of its income, it could not finance such large spending through internally generated capital. To remake itself, Western Union had to spend more money than it could afford.

While McFall's ventures into data processing and satellites looked flashy, customers continually complained of poor service in Western Union's core businesses. Marshall had maintained high standards of quality and timeliness for handling telegrams, if only to placate regulators and maintain the integrity of the Western Union brand, but McFall sought to kill off this low-margin business by continually raising rates and lengthening transmission times. By the late 1960s a domestic telegram cost more than a three-minute telephone call to Europe. In 1968 frustrated FCC commissioner Nicholas Johnson remarked that perhaps "the telegram, like the passenger ship and the carrier pigeon, has outlived its economic usefulness. But, if that be the case, a firing squad might be more humane than long lingering starvation."[72]

Despite its ambitious ventures into computers and satellites, Western Union remained a low-growth company. The story of Western Union's imminent turn-around had been a staple of the business press since the late 1940s, but the company's low growth and huge debt load made a lasting recovery difficult to achieve. Yet McFall continued spending. Between 1965 and 1971, the company doubled its capital investment, but its 4.4 percent return on equity was the lowest of all the country's major utilities. In 1972 the company's debt for capital investments amounted to just under $500 million, about equal to its equity, and the company had another $365 million in unfunded pension liabilities. By 1974 McFall had resorted to creative accounting procedures to hide the company's tightening financial straitjacket. That year Western Union reported a net profit to its shareholders, but its IRS tax return showed a net loss. Financial analysts discovered that the company was, in effect, using borrowed money to pay shareholder dividends and that expenditures consistently outpaced cash flow. Because analysts questioned the quality of the company's earnings, it was unable to raise money by issuing more stock or selling bonds except at high interest rates. Western Union relied on bank loans to finance its ongoing capital expansion programs. In a remark reminiscent of the Enron meltdown three decades later, one frustrated analyst complained, "You need a CPA, an attorney, a semanticist, and a metaphysician to understand what the footnotes in Western Union's reports mean."[73]

When McFall retired in 1979, he left a mixed legacy. He had rescued Western Union from technological obsolescence, but he had spent $2.5 billion to do it. The company's debt exceeded its equity. His successor Robert Flanagan sought to reduce expenditures by reorienting the company from hardware to services. He also cut staff, sold off unprofitable operations, and relied only on internally generated capital to finance development of three new and quite promising areas, AirFone telephones on board passenger airliners, cellular telephones, and email.[74]

With luck Flanagan might have succeeded in saving Western Union with a financially sound program of investment in these areas, but the financial situation bequeathed by McFall left no margin for the unexpected. Flanagan described his dilemma as like "turning an elephant around in a bathtub without spilling any water." In early 1984 Western Union lost a communications satellite shortly after launch. Although the $75 million satellite was insured for $100 million, it would have generated up to $500 million in revenue over its service life. At the end of the year, a consortium of eight banks refused to extend Western Union's credit, leaving it on the verge of bankruptcy. Two years later Western Union fell into the clutches of junk-bond dealer Drexel Burnham Lambert. One financial writer called the company "a huge, docile cash cow [that] had been dropped in [Drexel's] lap, ready to be carved up, hauled to market and sold in pieces for millions." Indeed, over the next few years, Western Union's management sold off all of its capital assets, its satellite network to Hughes, and its Telex network to AT&T. By 1990 all that remained of Western Union was its hugely profitable money transfer business. In 1994 this was sold along with the Western Union brand to First Data Corporation.[75]

The Final Telegram

Western Union sent its last telegram in January 2006. News stories reporting the event attributed the telegram's demise to developments since the 1980s, including deregulated long-distance telephone service, fax machines, and email, media that retain the rapidity or permanence of the telegram.[76] The popular conception of telegraphy is that it was a quintessential nineteenth- and twentieth-century medium overtaken by newer communication technologies in the past generation. True enough. Yet the story of the final telegram is also a story of the decline and fall of Western Union, a century-long story of shortsightedness, mismanagement, short-lived turnarounds and subsequent reversals, and, finally, dismemberment at the hands of 1980s junk-bond dealers. It is also the story of talented managers, who, from the 1940s onward, labored valiantly to save and rejuvenate the company. The task was ultimately impossible.

Western Union's recent history points up a problem often faced by business leaders but rarely noted in management literature. Most of the literature on corporate failure focuses on sins of commission or omission, blunders that executives made or shifts in technology or consumer preference that they failed to foresee.[77] Yet Western Union's postwar managers confronted an intractable problem: in the long run, the company might simply have been unsalvageable, yet its managers had to try.

As Western Union's history suggests, corporate failures become apparent and acute during crises, but they often have deep roots. Western Union failed in the 1980s because it could not resolve the dilemma of technological obsolescence on the one hand or overspending to modernize on the other. Yet this dilemma had persisted since World War II, and its roots reached back to its decision to cede the telephone to Bell in 1879. Adopting a longer historical perspective on Western Union's demise offers several lessons.

The most important lesson is that a company that succeeds through technological innovation and leadership must never lose that capacity. From the late 1860s to the mid-1870s Western Union fostered important new technologies like the ticker and quadruplex, and the company stood an excellent chance of driving the fledgling Bell company out of telephony. For Western Union, the appearance of the telephone was what Intel founder Andrew S. Grove has called a "strategic inflection point," a "time in the life of a business when its fundamentals are about to change. That change can mean an opportunity to rise to new heights. But it may just as likely signal the beginning of the end."[78]

It is clear that Western Union's managers did not understand that they confronted such a strategic inflection point in the late 1870s. However, charging them with failing to grasp the significance of the telephone is to blame them for failing to predict the future. Yet we can fault Western Union for its technological and organizational stagnation in the thirty years between the 1879 contract and AT&T's acquisition of control in 1909. Such conservatism is hardly surprising given Western Union's near-monopoly dominance of the telegraph market and the company's control in turn by the rival Vanderbilt and Gould interests.

Between 1870 and 1881, the telegraph industry was a secondary site of conflict between the Vanderbilts and Gould in their ongoing struggle to control the nation's railroads. The Vanderbilts before 1881, and Gould afterward, evinced little concern for Western Union's future and did not encourage its managers to think expansively about technological innovation or corporate mission.[79]

Thus, Western Union's leaders defined its market and mission narrowly, as a company that transmitted telegrams and stock quotations, and not as a network provider of electrical communication generally.[80] The telegraph industry's demise after World War II ultimately stemmed from this narrow definition of its mission. After the war, Western Union's managers sought to reclaim its stature in the communication field, and they undertook a crash program to do so. This very well might have succeeded if the company had had an internal research-and-development capability like most other major American corporations. This would have smoothed Western Union's transition from a telegram carrier to a modern telecommunications company. Instead, this transition was much more halting and expensive than it needed to be. When Marshall retired in 1965, the company, despite ambitious innovation and marketing programs, still depended on the declining telegram market for half its revenue.

Marshall's successor, Russell McFall, sought to reduce this dependence through a radical reorientation away from Western Union's core competencies and into new fields such as computers and satellites. McFall allowed the company's telegram service and rates to worsen, pushing the company further from its center of gravity and reducing the appeal of its brand. Under McFall, Western Union's debt ballooned until he ultimately bankrupted the company. Thus, a second lesson of Western Union's collapse is that such a radical transformation of a firm must be undertaken in a financially responsible way through a mix of internally generated and externally raised capital.

The intertwined histories of the telegraph and telephone industries also show that an integrated analysis of competing industries and government policies provides a more satisfying framework for understanding business success and failure. Limiting analysis to a particular firm or technology tends to produce a teleology in which success or failure seems inevitable. Yet the telephone did not supersede the telegraph simply because the former was an inherently superior technology or because Bell had better managers. Instead, legal, technical, and regulatory factors all interacted to form a porous and shifting boundary between the two industries. In this story regulatory considerations have been paramount. Antimonopoly sentiment, fear of hostile legislation, and a desire to avoid antitrust proceedings restrained Bell's managers from completely engulfing Western

Union from the 1880s onward. After the establishment of the FCC in 1934, the regulatory arena assumed even greater importance. This arena was not a neutral setting, and federal regulators were far from disinterested or aloof. Social and political issues were just as important as stark economic or technical consider-ations in determining the intertwined trajectories of the telegraph and telephone industries.[81]

A final lesson is that technologies often outlive their corporate shells. For more than a century, the telegram was synonymous with Western Union. At the turn of the twentieth century, perhaps 2 percent of Americans sent or received one telegram a year and the country as a whole sent an average of one per capita. In 2008, by contrast, there were an estimated 1.3 billion email users globally, each sending an average of 140 messages per day. In 2009 American mobile phone subscribers reportedly sent an average of fourteen text messages per day. While the once-ubiquitous telegraph key is now silent, the communication culture that it spawned continues to accelerate.[82]

The Promise of Telegraphy

In January 2010 Kodiak-Kenai Cable Company, an Alaskan telecommunications firm, announced plans to build an underwater fiber optic cable between London and Tokyo, a project now feasible because of the loss of summer sea ice in the Arctic. Although the project will cost more than $1 billion, its planners expect it to be profitable since it will cut the transmission time between the two cities from 140 to 88 milliseconds. This twentieth of a second difference, they contend, will draw traffic away from existing cables, because that small saving of time can translate into large differences in the profitability of financial transactions.[1]

In one sense this project is the logical culmination of the electrical communication revolution that began more than 150 years ago. A common refrain during the nineteenth century was that the telegraph promised to annihilate space and time. Samuel Morse, seeking funding to build his first experimental line, told Congress in 1838 that his invention would "diffuse with the speed of thought, a knowledge of all that is occurring throughout the land, making in fact one *neighborhood* of the whole country." The House Committee on Commerce, chaired by Morse's business partner, Francis O. J. Smith, agreed, claiming that "the citizen will be invested with, and reduce to daily and familiar use... the HIGH ATTRIBUTE OF UBIQUITY" formerly reserved to the divine. Thus, Morse's telegraph promised "a revolution unsurpassed in moral grandeur" by any other past invention.[2]

The most famous counterexample to this utopian line of thought was the iconoclast Henry David Thoreau. In a celebrated remark, he labeled technologies like the railroad and telegraph "improved means to an unimproved end." Instead of placing the parts of the country into spiritual communion with each other, he thought it likely that "Maine and Texas . . . have nothing important to communicate." Even a projected Atlantic cable, an engineering undertaking of mammoth proportions for the day, might convey only the news that "Princess Adelaide has the whooping cough." Yet Thoreau never questioned the fundamental premise of technological utopians, that the telegraph made possible easy and instant communication across long distances, even if the messages themselves might be trivial.[3]

Morse's and Smith's enthusiastic (if self-serving) pronouncements and Thoreau's critique help connect the potential of telegraphy with the realities of its ownership, control, and use. The issue at stake is that electrical communications technology holds out the promise of nearly frictionless information transmission, if one neglects issues of cost, access, and message content. This promise is worth some reflection, even if it is closer to reality on today's fiber-optics links than on nineteenth-century telegraph wires.[4]

Of course, no technology lives up to the expectations of its most sanguine promoters and enthusiasts. However, the gap between telegraphy's theoretical promise and actual performance was of vital concern to telegraph managers, reformers, and customers. The main questions for them were who ought the telegraph serve and for what ends? The answers to these questions depended on the attributes of the technology itself as well as the social contexts within which the telegraph operated.

Consider first the capabilities and limitations of the technology itself. The telegraph promised instant communications across long distances because electrical signals travel through wires nearly instantly. Thus, the telegraph provided rapid office-to-office communication through a geographically extended network. However, telegraphy could not give users the immediacy that early enthusiasts had hoped for. Instead, telegrams were mediated forms of communication, requiring sending and receiving operators and messengers for delivery. Not only did this structure make secrecy and dialogue difficult, but it also meant that the delivery of a telegram could take anywhere from ten minutes to several hours. Furthermore, because it was capital intensive to build telegraph lines and labor intensive to operate them, the industry tended toward concentration, under either private control or state ownership.

In the United States, Western Union's managers answered the questions of whom their network ought to serve and for what purposes narrowly. Between 1866 and 1909, they consistently rejected any public-service ethos. They regarded the telegraph as a communications system primarily for the press, brokers, and businessmen. Ordinary Americans did not use the telegraph, they asserted, because it was an expensive medium best suited for the transmission of urgent and important messages, news dispatches, and market quotations. This narrow vision of telegraphy's social roles stunted Western Union's business strategies regarding technological development and marketing and its response to evolving politico-economic thought about public utilities.

Postal telegraph reformers answered these questions very differently. They grasped the possibilities of telegraphy as a popular, social, and educational medium. However, they never fully acknowledged the limitations of their vision. In particular, reformers, including Gardiner Hubbard and the Washburn(e) brothers, failed to understand the economics of the technology. They relied on overly optimistic assumptions to claim that government operation would make telegraphing affordable for the million while improving telegraphers' wages and working conditions, all without incurring deficits for the government. Others, such as public-ownership advocate Frank Parsons and British telegraph engineer William Preece, acknowledged that deficits were likely, but they argued that the social benefits of government operation outweighed the financial losses. When put to the test during World War I, postal management turned out to be disastrous, ending the postal telegraph movement for good and eroding support for public ownership of utilities generally.

The British nationalization of 1870 was a main source of inspiration to American reformers. The British experience left a mixed legacy. On the positive side, the Post Office telegraphs popularized telegraphy and made it more affordable. Most Continental countries (the major exceptions being Switzerland, the Netherlands, and Belgium) regarded the telegraph primarily as a medium for official government messages and only secondarily as a means of private communication. By 1875, Britons sent nearly three times as many telegrams per capita as the French and Germans, with average message length about double those sent by their Continental counterparts. The British Post Office also aided the development of the provincial press. Before nationalization, newspapers outside London paid high rates for news dispatches. This was a major reason why most of the British press supported nationalization, in contrast to the opposition of news-brokers in the United States. However, the British postal telegraph consistently ran deficits, partly because of the inflated cost of buying out the private telegraph companies but also owing to the low rates charged newspapers for telegraphic news. Also, government ownership of the telegraphs retarded the technological development and market diffusion of a competing technology, the telephone, by about a generation.[5]

A final area of comparison linking the technical and economic attributes of telegraphy to business and regulatory contexts is the relationship between the telegraph and financial markets. In the United States, ticker service and private wire leases came the closest to fulfilling Francis O. J. Smith's vision of ubiquity, at least within the financial world. Together, these systems made space irrelevant for market transactions, thus modernizing American financial markets and broadening participation in them. In Europe, however, ticker service was far less important to the development of financial markets. While the London Stock Exchange had a limited ticker service, it did not provide continuous quotations as American exchanges did. The Paris Bourse was a government institution, with only seventy *agents de change* licensed to buy and sell stocks on its floor. Germany placed a stamp tax on stock sales that damped growth of its securities markets. Thus, most European brokers saw no need to install tickers in their offices. Furthermore, the bucket shop was a purely American phenomenon because government-run telegraph systems restricted transmission of financial information to legitimate brokers. Because of these infrastructural, institutional, and legal differences, American participation in financial markets remained broader than in Europe for most of the twentieth century.[6]

During the early years of the telegraph, technological utopians like Morse and Smith had the luxury of considering the technology in isolation from social, economic, and political contexts. An immature technology contains vast possibilities for social transformation. Only after a technology is placed into these contexts can we see which possibilities are enhanced and diminished. We can draw several conclusions from this.

To begin with, a technology does not effect change by itself but requires mediation through existing institutions and often leads to the creation of new ones. For example, the ticker played different roles in the development of American and European financial markets, mainly because of differences in how those markets operated. In the United States, the ticker helped create a new institution, the bucket shop, through which the speculating public gained acquaintance with the mechanics of stock trading.

The social impact of the telegraph was not limited to institutions. Telegraphy also changed how individuals understood themselves in relation to these institutions. To fully understand the telegraph's impact on American life, we must pay greater attention to the psychological dimensions of technological change.[7] The way in which ordinary Americans responded to frequent telegraphic updates of important news stories during and after the Civil War is a salient example of these psychological effects. We must also consider how individual responses and choices interact with technological applications. For example, investors and speculators eagerly adopted the stock ticker after about 1870, making it one of the fastest growing and most profitable segments of the telegraph industry. But the ticker also changed how traders regarded financial markets, transforming them from places to trade tangible goods to the continuous flow of quotations printed on the tape and posted on brokers' blackboards. The rise of modern finance capitalism owed as much to the evolving psychology of market participants as it did to institutional changes wrought by telegraphy.

Centralizing and decentering tendencies are both inherent in communication technologies. The United States Military Telegraph allowed civilian officials to exercise unprecedented oversight of military affairs. At the same time, the wartime telegraph gave field commanders greater operational flexibility and fostered a sense of craft autonomy among telegraph operators. Similarly, the twin monopoly of Western Union and the Associated Press led to centralized control over the production and distribution of news. However, telegraphic news reporting gave more Americans access to news than ever before.

The differential social impact of telegraphy helps us to understand the historical relationship between communication and information. Although the Bell System succeeded in making the telephone a social medium,[8] Western Union was unwilling to do so for the telegraph until about 1910. Afterward, when Western Union's managers did try to popularize telegrams, they found that most Americans had come to regard them as bearers of bad news. This marketing problem dogged the company until the telegraph's ultimate obsolescence. For most Americans, the telegraph was less a communication medium than a source of broadcasted information like wire-service news stories and stock and commodity quotations. Thus, it was as a medium for the distribution of information rather than as a communications medium that the socially transformative nature of telegraphy becomes readily apparent.

Finally, the line of development joining Morse's telegraph to the projected fiber optic link under the Arctic prompts deeper reflection on the relationship between technology and social change. To claim that the telegraph was socially transformative is to raise the issue of technological determinism. Historians of technology generally claim that the field has outgrown technological determinism as an explanatory framework. However, we must acknowledge that technology shapes society. This recognition need not imply that technology is the only, the most important, or an inevitable cause of social change. After all, historical events and trends have complex causes and antecedents. A satisfying analysis of the relationship between technology and society must preserve individual agency and remain sensitive to social context while fully acknowledging that a technology's material attributes and capabilities shape its effects on individuals and societies. Historians of technology are uniquely positioned to show how technologically driven social change has become fundamental to the human experience.[9]

1794
The Chappe brothers build an optical telegraph between Paris and Lille for the exclusive use of the French government. By the 1840s, the network connects all major French towns.

1800
Italian scientist Alessandro Volta invents an electric battery, the first device able to provide a continuous electric current. Previously, electricians used static-electricity machines that were incapable of producing continuous current.

1801
Jonathan Grout builds an optical telegraph in Boston harbor. The line draws few customers and closes in 1807.

1820
Danish scientist Hans Christian Oersted discovers electromagnetism when he observes the deflection of a compass needle while he opens and closes a nearby electric circuit.

1822
John Rowe Parker sets up an optical telegraph in Boston harbor. It remains in operation until the 1850s.

1823
Captain Samuel Reid builds an optical telegraph connecting New York City, Staten Island, and Sandy Hook, N.J. It remains in service until at least 1856.

1824
British scientist William Sturgeon invents the electromagnet by coiling wire around a piece of iron. His electromagnet is the first device able to convert electricity into mechanical work.

1830
American physicist Joseph Henry develops greatly improved electromagnets. One design can lift hundreds of pounds of weight and the other can be actuated at great distances from a battery. Henry later demonstrates that his second type of electromagnet can perform simple signaling over long distances by tapping a small bell through several miles of wire.

OCT. 1832
Samuel F. B. Morse conceives of an electromagnetic recording telegraph during an Atlantic crossing. His design imprints dots and dashes onto a moving strip of paper.

JAN. 1836
Morse demonstrates a crude prototype of his telegraph to chemist Leonard D. Gale, who urges Morse to adopt Joseph Henry's improved long-distance electromagnet. Morse does so and the range of his telegraph jumps from forty to seventeen hundred feet. Morse later gives Gale a share of his telegraph patent.

JAN. 1837
Captain Samuel Reid urges Congress to build an optical telegraph along the eastern seaboard.

SEPT. 1837
Morse hires Alfred Vail to refine the mechanical operation of his telegraph for demonstration purposes in exchange for a share of a future telegraph patent.

SEPT. 1837
Morse files a caveat for a patent on his telegraph.

FEB.–MAR. 1838
Morse demonstrates his telegraph to the Franklin Institute and to President Van Buren and his cabinet.

APR. 1838
Maine congressman Francis O. J. Smith sponsors a bill to appropriate $30,000 for Morse to build a demonstration telegraph line. The bill fails. Smith conceals that he had recently become a partner of Morse, Gale, and Vail.

MAY 1838
British inventors William Fothergill Cooke and Charles Wheatstone install their electromagnetic telegraph on a British railroad. Their design differs from Morse's because it does not record signs but uses magnetized needles to point to letters on a board.

1840
Morse obtains a patent for his electromagnetic recording telegraph.

MAR. 1843
After five years of effort, Morse succeeds in getting a congressional appropriation to build a demonstration line from Washington to Baltimore.

MAY 1844
Morse opens his demonstration line from Washington to Baltimore. Although he

lobbies Congress to buy his patent rights for a few more years, Congress does not.

MAR. 1845
Morse and Vail hire Amos Kendall, who had been President Jackson's postmaster general, as their business manager.

MAY 1845
Kendall incorporates the Magnetic Telegraph Company, the first privately owned telegraph line, to operate between Washington and New York.

1845
Vermont inventor Royal E. House designs a telegraph that prints Roman letters onto a moving strip of paper. He patents it a year later. Although it has some early commercial success, it proves unreliable and falls out of use by the late 1850s.

1846
Morse patents an electromagnetic relay.

MAR. 1847
Magnetic Telegraph buys the demonstration Washington-to-Baltimore line from the Post Office Department.

1848
Six New York newspapers form the New York Associated Press (NYAP) to share the costs of obtaining telegraphic news, formalizing arrangements for telegraphic newsgathering that had been in place since 1846.

1849
Scottish inventor Alexander Bain obtains a U.S. patent for a telegraph design that imprints dots and dashes onto chemically treated paper. Although Bain's telegraph does not use electromagnetism, the Morse interests win a judgment that it infringes Morse's patent.

1854
The U.S. Supreme Court upholds Morse's patent but strikes down its broad claim to

any telegraph design using electromagnetism, confining it to just his particular design. However, the House and Bain telegraphs prove unreliable in service and within a few years the U.S. telegraph industry uses Morse's design almost exclusively.

1856
Hiram Sibley of Rochester, N.Y., merges several regional telegraph companies to form the Western Union Telegraph Company.

1856
New York investors form the American Telegraph Company to link the American seaboard with prospective landing sites of a planned Atlantic cable in Canada. By 1859 it controls all telegraph traffic along the Atlantic coast of the United States from the Canadian border to New Orleans.

1857
The six major telegraph companies (Western Union; American; Illinois and Mississippi; South Western; Atlantic and Ohio; and New York, Albany, and Buffalo) form the North American Telegraph Association, a cartel to divide up the telegraph market geographically and to block new competitors. The Montreal Telegraph Company joins in 1858. The agreement does not include telegraph companies controlled by the Morse patentees, which merge with the American Telegraph Company in 1859.

AUG. 1857
A joint British-American expedition unsuccessfully attempts to lay an Atlantic cable.

AUG. 1858
Another expedition succeeds in laying an Atlantic cable between Valentia, Ireland, and Heart's Content, Newfoundland. It works for a few weeks before failing.

JUNE 1861
Morse's 1840 telegraph patent expires. Originally set to run fourteen years, Morse had received a seven-year extension.

JULY 1861
The federal government begins censoring telegraphic news from Washington.

OCT. 1861
The Pacific Telegraph Company completes a transcontinental telegraph line to California. Although a nominally separate company, it is really under the control of Western Union.

OCT. 1861
The War Department sets up the U.S. Military Telegraph (USMT).

JAN.–MAR. 1864
Western Union acquires the New York, Albany, and Buffalo and Atlantic and Ohio Telegraph companies.

1864
Emboldened by repeated failures to lay a working Atlantic cable, Western Union begins building an overland telegraph line to link the hemispheres. The line is to run north along the Pacific coast from California to Alaska, through an underwater cable across the Bering Strait, and through Siberia to connect to the European telegraph network in Russia.

AUG. 1864
United States Telegraph Company is formed.

AUG. 1865
Another attempt to lay an Atlantic cable fails.

NOV. 1865
William Orton becomes president of United States Telegraph Company.

JAN. 1866
American Telegraph acquires South Western Telegraph Company.

FEB. 1866
War Department disbands USMT and turns over its lines to the commercial telegraph companies, though some officers are not mustered out until the summer.

FEB. 1866
Western Union acquires United States Telegraph. William Orton becomes a Western Union vice-president.

APR. 1866
War Department ends telegraphic censorship.

JUNE 1866
Western Union acquires American Telegraph, its last major rival.

JULY 1866
Congress passes the National Telegraph Act. The law gives telegraph companies agreeing to its provisions access to certain railroad rights-of-way in exchange for transmitting government messages at rates set by the postmaster general. The law also sets up a method for Congress to buy out telegraph companies that assent to the act's provisions at any time after the law has been in effect for five years.

AUG. 1866
A permanently working Atlantic cable begins operation between Newfoundland and Ireland.

1866
Samuel Spahr Laws invents a gold indicator that instantly reports results of trades on the New York Gold Exchange.

JAN. 1867
Western Union forces a settlement between the NYAP and regional press associations, especially the Western Associated Press (WAP), that weakens NYAP's power over them. The major conflict is over cost sharing for news transmitted through the Atlantic cable. An important provision of this agreement is that members of NYAP

and regional press associations agree not to support any telegraph line competing with Western Union, including a postal telegraph.

MAR. 1867
Western Union halts construction of the Russian-American telegraph project since it is apparent that the Atlantic cable will work permanently.

JUNE 1867
Western Union accepts the provisions of the National Telegraph Act.

JUNE 1867
Illinois and Mississippi Telegraph Company, the last holdout of the North American Telegraph Association, leases it lines to Western Union. In May 1868 Western Union acquires it outright.

JULY 1867
William Orton succeeds Jeptha Wade as Western Union president.

1867
Edward A. Calahan invents the stock ticker, a machine that prints the results of stock and commodity trades onto a moving strip of paper.

1868
Joseph Stearns, president of Franklin Telegraph Company, invents a practical duplex telegraph capable of sending two messages simultaneously (one in each direction) over a telegraph line.

1870
The British Post Office assumes control of the previously privately owned inland telegraphs.

1870
The Army Signal Service inaugurates a weather reporting and prediction system using the lines of commercial telegraph companies.

1870

Control of Western Union shifts from its original group of Rochester, N.Y., investors headed by Hiram Sibley to a New York City group headed by railroad tycoon Cornelius Vanderbilt.

MAR.–MAY 1871

Western Union refuses to handle Army Signal Service weather reports because of a dispute over billing and circuit usage.

MAY 1871

Western Union acquires control of the Gold and Stock Telegraph Company, the leading supplier of market quotations.

DEC. 1871

President Ulysses Grant urges Congress to buy out Western Union per the provisions of the National Telegraph Act.

1872

Joseph Stearns sells the patent rights for his duplex to Western Union.

JULY 1872–FEB. 1874

Western Union again refuses to handle Signal Service weather reports.

1874

Jay Gould acquires control of the Atlantic and Pacific Telegraph Company, which begins a fierce competition with Western Union. This marks Gould's first attempt to gain control of Western Union.

1874

Edison invents the quadruplex, capable of sending four simultaneous telegraph messages, two in each direction, on one telegraph line. Although Edison did this work for Western Union, that company's tardiness to compensate him leads him to sell his quadruplex rights to Gould's Atlantic and Pacific Telegraph. The resulting dispute is settled in late 1875 when Edison assigns his rights to Western Union.

FEB. 1875

Alexander Graham Bell demonstrates his harmonic telegraph, theoretically capable of transmitting several telegraph messages simultaneously, to Western Union officials.

14 FEB. 1876

Alexander Graham Bell files a patent for the electrical transmission of sound, the basis of his telephone design. On the same day, Elisha Gray files a patent caveat for a similar idea. Bell develops a primitive telephone prototype the following month.

APR. 1877

The Bell telephone company leases its first telephones.

APR.–JUNE 1877

British Post Office telegraph engineers Henry Fischer and William Preece travel to the United States to study its telegraph system.

OCT. 1877

Atlantic and Pacific and Western Union sign a rate-pooling agreement, effectively ending competition between them.

DEC. 1877

Western Union sets up the American Speaking Telephone Company, a subsidiary intended to compete with Bell in the telephone market. American Speaking Telephone uses the telephone patents of Thomas Edison and Elisha Gray.

1877

The first bucket shops, betting parlors where patrons wagered on ticker quotations of stocks and commodities, appear in New York City.

APR. 1878

William Orton dies. Norvin Green succeeds him as Western Union president.

MAY 1879
Jay Gould sets up American Union Telegraph Company to compete with Western Union, his second attempt to gain control of Western Union.

JUNE 1879
Congress passes an amendment to an army appropriations bill sponsored by Massachusetts congressman Benjamin F. Butler. The Butler amendment allows railroads assenting to the terms of the National Telegraph Act to operate a public telegraph business and helps Gould build the American Union into a national telegraph company.

JULY 1879
Western Union begins leasing telegraph circuits to private parties. Its first lease is to brokerage Edward K. Willard & Company for a line from New York City to Saratoga, N.Y.

NOV. 1879
Western Union and the Bell telephone interests sign a contract partitioning the electrical communication industry. Western Union agrees to exit the telephone industry in exchange for Bell's commitment not to enter the telegraph market.

JAN. 1881
Jay Gould succeeds in obtaining control of Western Union from the Vanderbilt interests. Shortly afterward, Western Union buys in the Atlantic and Pacific and American Union companies. Gould's acquisition of Western Union reanimates the postal telegraph movement.

DEC. 1882
Western Union forces a settlement between NYAP and WAP, installing WAP general manager William Henry Smith as head of both organizations.

JULY–AUG. 1883
Telegraph operators launch a strike against Western Union. Although unsuccessful, they gain much public sympathy. The strike increases support for a postal telegraph.

1883
Wealthy Nevada silver miner John W. Mackay buys control of the Postal Telegraph Company and sets up an Atlantic cable company with *New York Herald* publisher James Gordon Bennett Jr. Postal becomes Western Union's longest-surviving competitor.

1886
Bell sets up its first long-distance telephone line between New York and Philadelphia. In early 1887 it begins leasing private long-distance circuits for both telephony and telegraphy.

1887–94
Economist Richard T. Ely popularizes the concept that telegraph companies are natural monopolies and therefore ought to be publicly owned.

1890
Bell begins using Belgian inventor François Van Rysselberghe's system for transmitting voice and telegraph traffic simultaneously on circuits it leases to private customers.

JUNE 1891
The U.S. Army turns over weather reporting to the Department of Agriculture and disbands the Signal Service. Henceforth, the army's Signal Corps restricts itself to military communication.

FEB. 1893
Norvin Green dies, and Thomas T. Eckert succeeds him as Western Union president.

1893
Western Union begins replacing its batteries with dynamos to provide power to its telegraph network.

1900
WAP recharters in New York as the Associated Press.

MAR. 1902
Thomas T. Eckert retires as Western Union's president. Robert C. Clowry succeeds him.

1904
The U.S. Supreme Court upholds telegraph companies' exclusive rights-of-way contracts with railroads in *Western Union Telegraph Company v. Pennsylvania Railroad Company*.

1904
Under pressure from Jay Gould's daughter Helen, a major stockholder, Western Union ceases supplying horse-racing results directly to gamblers and bookmakers.

1905
The U.S. Supreme Court rules in *Board of Trade v. Christie Grain & Stock Co.* that stock and commodity exchanges own the quotations generated on their floors. This ruling is the legal basis for the eradication of bucket shops over the next decade.

1907
Theodore N. Vail becomes president of AT&T.

AUG.–NOV. 1907
Commercial Telegraphers Union of America launches an unsuccessful strike against Western Union.

1909
Under the leadership of Theodore N. Vail, Bell acquires working control of Western Union. Vail proceeds to set up a "universal wire system" combining the telegraph and telephone networks.

JUNE 1910
Congress passes the Mann-Elkins Act, placing the telegraph industry under In-terstate Commerce Commission (ICC) jurisdiction. ICC regulation is light.

DEC. 1913
Bell agrees to divest itself of its control of Western Union under threat of antitrust prosecution. It relinquishes control by the end of 1914. AT&T and Western Union agree to pool their current and future patents related to telegraph printers; this agreement does not include patents related to transmission methods.

JULY 1918
President Woodrow Wilson claims wartime necessity to place the telegraph and telephone networks under Post Office control.

16 NOV. 1918
President Wilson transfers control of overseas cables landing on U.S. soil to the Post Office Department, despite the cessation of hostilities on 11 November.

1918
AT&T installs the first commercial carrier-current transmission system on a long-distance telephone circuit between Baltimore and Pittsburgh.

MAY 1919
Congress returns control of overseas cables to their private owners.

JULY 1919
Congress returns control of the telephone and telegraph networks to their private owners.

1924
Western Union installs a submarine cable using inductive loading between New York and the Azores. Largely a product of AT&T's research-and-development laboratories, this cable increases transmission speed on ocean cables by about a factor of ten.

1928
AT&T begins leasing spare circuit capacity to Western Union. It is able to do so because of its control over carrier-current transmission technology.

1931
AT&T inaugurates a successful Teletypewriter Exchange service (TWX) that makes serious inroads into Western Union's leased-wire business.

1934
Federal Communications Commission (FCC) is established with jurisdiction over radio, overseas cables, telegraphs, and telephones.

1935
Western Union sets up a prototype facsimile circuit between New York and Buffalo.

1940
Western Union perfects facsimile transmission of telegrams, although full commercial development does not occur until after World War II.

1943
At the behest of the FCC, Western Union reluctantly takes over a nearly bankrupt Postal Telegraph Company.

1945
Western Union begins construction of a nationwide microwave transmission network using carrier-current transmission to replace telegraph wires.

1948
Walter P. Marshall becomes president of Western Union.

1956
TAT1, the first Atlantic cable for telephone circuits, enters service. It is a joint project of AT&T and the British and Canadian governments.

1958
Western Union rolls out its Telex service that allows direct customer-to-customer communication using self-dialing.

1965
Russell McFall succeeds Walter P. Marshall as president of Western Union.

1974
Western Union launches Westar 1, a geosynchronous communications satellite. Between 1974 and 1984 Western Union launches seven communications satellites.

1979
Robert Flanagan succeeds Russell McFall as president of Western Union.

1984
Failure of a satellite launch and inability to secure a line of credit leaves Western Union on the verge of bankruptcy.

1986
Drexel Burnham Lambert Incorporated takes over Western Union and begins selling off its assets.

1994
Western Union sells its money-transfer business and brand to First Data Corporation.

2006
Western Union sends its last telegram.

Abbreviations

A&P Annual Report   *Annual Report of the Atlantic & Pacific Telegraph Company* (New York: Van Kleeck, Clark, various years)

BoD CBOT   Board of Directors Meeting Documents, Chicago Board of Trade Records, Special Collections, University of Illinois at Chicago Library

CBOT   Chicago Board of Trade Records, Special Collections, University of Illinois at Chicago Library

CSO Annual Report   *Annual Report of the Chief Signal-Officer to the Secretary of War* (Washington, D.C.: GPO, various years)

G&S   Gold and Stock Telegraph Company

G&S Annual Report   Gold and Stock Telegraph Company Annual Report, Western Union Telegraph Company Records, Lemelson Center, Smithsonian Institution, Washington, D.C..

G&S BoD   Board of Directors Minutes, Gold and Stock Telegraph Company, Western Union Telegraph Company Records, Lemelson Center, Smithsonian Institution, Washington, D.C.

G&S ExCom   Executive Committee, Gold and Stock Telegraph Company, Western Union Telegraph Company Records, Lemelson Center, Smithsonian Institution, Washington, D.C.

Misc. Docs., CBOT   Miscellaneous documents, Chicago Board of Trade Records, Special Collections, University of Illinois at Chicago Library

NYSE   New York Stock Exchange Archives, New York

P&A Annual Report   *Annual Report of the President, Secretary and Treasurer of the Pacific & Atlantic Telegraph Co. of the United States* (Pittsburgh: W. G. Johnston, various years)

PLB WUTC	President's Letterbooks, Western Union Telegraph Company Records, Lemelson Center, Smithsonian Institution, Washington, D.C.
Statistical Notebooks, WUTC	Statistical Notebooks, 1886–1908, Western Union Telegraph Company Records, Lemelson Center, Smithsonian Institution, Washington, D.C.
TAE Digital Edition	Thomas A. Edison Papers Digital Edition, http://edison .rutgers.edu/digital.htm
TAE Papers	The Papers of Thomas A. Edison (Baltimore: Johns Hopkins University Press, 1989–2011)
USAG Annual Report	*Annual Report of the Attorney-General of the United States* (Washington, D.C.: GPO, various years)
WUTC	Western Union Telegraph Company Records, Lemelson Center, Smithsonian Institution, Washington, D.C.
WUTC Annual Report	*Annual Report of the President of the Western Union Telegraph Company* (New York: Russell Brothers, various years)
WWP	Arthur S. Link, ed., *The Papers of Woodrow Wilson* (Princeton: Princeton University Press, 1966–94.

Introduction. Why the Telegraph Was Revolutionary

1. "Petition of Samuel C. Reid, Praying the Establishment of a Line of Telegraphs from New York to New Orleans," 26 Jan. 1837, Committee on Post Office and Post Roads, H.R. Doc. 107, 24th Cong., 2d sess., *Congressional Globe*, vol. 30, 24th Cong., 2d sess., 3 Feb. 1837, 151; "Letter from the Secretary of the Treasury, Transmitting a Report upon the Subject of a System of Telegraphs for the United States," 11 Dec. 1837, H.R. Doc. 15, 25th Cong., 2d sess. On Reid, see Reginald Horsman, "Reid, Samuel Chester," *American National Biography Online*, www.anb.org, accessed 25 Mar. 2012.

2. Gerard J. Holzmann and Bjorn Pehrson, *The Early History of Data Networks* (New York: Wiley Interscience, 2003); Geoffrey Wilson, *The Old Telegraphs* (London: Phillimore, 1976); Alvin Harlow, *Old Wires and New Waves: The History of the Telegraph, Telephone, and Wireless* (New York: D. Appleton-Century, 1936), 13–34; Robert Greenhalgh Albion, *The Rise of New York Port, 1815–1860* (New York: Charles Scribner's Sons, 1939), 217–20; I. N. Phelps Stokes, *The Iconography of Manhattan Island* (New York: Robert H. Todd, 1909–28; repr., New York: Arno Press, 1967), vol. 5, entries for 12 Mar., 12 June, 25 June, 30 June 1812; 13 Mar., 14 May, 3 July, 26 July, 11 Sept. 1813; 17 Jan. and 12 Aug. 1816; 11 and 23 June 1821; 2 May 1827; John R. Parker, "Marine Telegraph System," *Army and Navy Chronicle* 11 (20 Aug. 1840): 122.

3. Morse gave the first public demonstration of his telegraph, through a circuit of seventeen hundred feet, in his rooms at New York University on 2 Sept. 1837. His most recent biographer Kenneth Silverman notes that his "response to Woodbury reconceived the whole project." Silverman, *Lightning Man: The Accursed Life of Samuel F. B. Morse* (New York: Alfred A. Knopf, 2003), 160.

4. When referring to *technological practice*, I have in mind Edward Constant's concept of technological revolution, akin to Thomas Kuhn's scientific revolution, as a paradigm shift in technical practice. Constant, *The Origins of the Turbojet Revolution* (Baltimore: Johns Hopkins University Press, 1980).

5. Paul Israel, *From Machine Shop to Industrial Laboratory: Telegraphy and the Changing Context of American Invention, 1830–1920* (Baltimore: Johns Hopkins University Press, 1992); David R. Meyer, *Networked Machinists: High-Technology Industries in Antebellum America* (Baltimore: Johns Hopkins University Press, 2006).

6. For an extended discussion of this point, see David Hochfelder, "Taming the Lightning: American Telegraphy as a Revolutionary Technology, 1832–1860" (Ph.D. diss., Case Western Reserve University, 1999), 137–45. See also Ross Thomson, *Structures of Change in the Mechanical Age: Technological Innovation in the United States, 1790–1865* (Baltimore: Johns Hopkins University Press, 2009), 173, 180, 246–52, 282–83.

7. I am drawing on the insights of W. Brian Arthur, *The Nature of Technology: What It Is and How It Evolves* (New York: Free Press, 2009), 156–59.

8. Although it had been in commercial operation since 1844, the telegraph industry did not provide customers reliable service or earn their trust until the late 1850s, after the formation of a cartel of six major companies that established standard accounting and message-handling procedures. Robert Luther Thompson, *Wiring a Continent: The History of the Telegraph Industry in the United States, 1832–1866* (Princeton: Princeton University Press, 1947).

Chapter 1. "Here the Telegraph Came Forceably into Play":
The Telegraph during the Civil War

1. Bound vol. 11, Personal Journal, William L. Gross Papers, Western Reserve Historical Society, Cleveland, Ohio.

2. William R. Plum, *The Military Telegraph during the Civil War in the United States* (Chicago: Jansen, McClurg, 1882), 2:349–50; Robert Luther Thompson, *Wiring a Continent: The History of the Telegraph Industry in the United States, 1832–1866* (Princeton: Princeton University Press, 1947), 394–95.

3. On the Confederate telegraphs, see J. Cutler Andrews, "The Southern Telegraph Company, 1861–1865: A Chapter in the History of Wartime Communications," *Journal of Southern History* 30 (Aug. 1964): 319–44. On the North's superior harnessing of the railroads and telegraphs for logistics purposes, see Edward Hagerman, *The American Civil War and the Origins of Modern Warfare: Ideas, Organization, and Field Command* (Bloomington: Indiana University Press, 1988).

4. David Homer Bates, *Lincoln in the Telegraph Office: Recollections of the United States Military Telegraph Corps during the Civil War* (New York: Century, 1907). Stanton's "right arm" quote is on p. 11 and is also in Plum, *Military Telegraph*, 2:324. For examples of Seward's advance reports of European news, see Edward S. Sanford to Seward, 26 Jan., 11 Feb., and 24 Nov. 1862, William H. Seward Papers, Manuscript Division, Library of Congress, Washington, D.C.

5. Amos Kendall to John Caton, 14 Mar. 1861, John Dean Caton Papers, Manuscript Division, Library of Congress, Washington, D.C.

6. Doris Kearns Goodwin, *Team of Rivals: The Political Genius of Abraham Lincoln* (New York: Simon and Schuster, 2007), 352–53.

7. "Report of the Secretary of War to President Lincoln, July 1, 1861," in Appendix to *Congressional Globe*, 37th Cong., 1st sess., 11–12; Thompson, *Wiring a Continent*, 384–85; Donald E. Markle, ed., *The Telegraph Goes to War: The Personal Diary of David Homer Bates, Lincoln's Telegraph Operator* (Lynchburg, Va.: Schroeder Publications, 2005), 23–24.

8. Plum, *Military Telegraph*, 1:93. For copies of governors' orders authorizing Stager to assume control of telegraph lines, see McClellan to Gov. Yates of Illinois, 13 May 1861, and order of Gov. Oliver P. Morton of Indiana, 14 May 1861, both in Caton Papers.

9. Plum, *Military Telegraph*, 1:68, 91–132.

10. Stephen B. Sears, ed., *The Civil War Papers of George B. McClellan: Selected Correspondence, 1860–1865* (New York: Da Capo, 1992), 123.

11. Ibid., 40–42, 79, 107, 181–82; Thompson, *Wiring a Continent*, 387; Charles A. Dana, *Recollections of the Civil War: With the Leaders at Washington and in the Field in the Sixties* (New York: D. Appleton, 1913), 9–10; Ulysses S. Grant, *Memoirs and Selected Letters* (New York: Library of America, 1990), 403.

12. Dana, *Recollections of the Civil War*, 112.

13. 12 May 1864, Luther A. Rose Diary, Manuscript Division, Library of Congress, Washington, D.C.; Plum, *Military Telegraph*, 2:134–35.

14. 12 and 18 May, 20 June, and 29–31 July 1864, Rose Diary; Plum, *Military Telegraph*, 2:259. For details of the Wilderness and Crater, see David J. Eicher, *The Longest Night: A Military History of the Civil War* (New York: Simon & Schuster, 2008), 676–78, 720–23.

15. Quartermaster General Montgomery C. Meigs Report to Secretary of War Stanton, 3 Nov. 1864, for period ending 30 June, in *The War of the Rebellion: A Compilation of the Official Records of the Union and Confederate Armies* (Washington, D.C.: GPO, 1880–1901), ser. 3, 4:885–86; George B. McClellan, *McClellan's Own Story: The War for the Union, the Soldiers Who Fought It, the Civilians Who Directed It, and His Relations to It and to Them* (New York: Charles L. Webster, 1887), 54, 134–35; William Tecumseh Sherman, *Memoirs of Gen. W. T. Sherman, Written by Himself*, 4th ed. (New York: Charles L. Webster, 1891), 2:398; Ulysses S. Grant, *Personal Memoirs of U. S. Grant* (New York: Library of America, 1990), 534–36; "Re-Union Proceedings of the Society of the United States Military Telegraph Corps. Chicago, September 19–20, 1883"; and "Sixth Reunion Proceedings of the Society of the United States Military Telegraph Corps." The New York Public Library holds the "Reunion Proceedings of the Society of the United States Military Telegraph Corps" from 1882 to 1898.

16. Roscoe Pound, "The Military Telegraph in the Civil War," *Proceedings of the Massachusetts Historical Society* 66 (1942): 185–203.

17. Cases of John W. Rockwell, J. D. Flynn and Charles K. Hambright, Michael

Deslonde, Anthony R. Walsh, Michael Mansfield, and W. C. McReynolds, Records of the Judge Advocate General, U.S. Army, Court Martial Case Files, RG 153, National Archives, Washington, D.C. An alphabetical index to the court-martial files is in the Finding Aids room at the National Archives; the index also contains the occupation or unit of the subjects.

18. Entries for 6, 22, 24 Oct. and 11 Nov. 1862, "Official Records, Telegraph Office, Cairo, Ill., Aug. 15, 1862 to Dec. 22, 1862," contents of box 12, folder 1, relating to a threatened strike of nineteen operators, Gross Papers; Plum, *Military Telegraph*, 1:265–70, 2:113–16, and 2:170–74.

19. Plum, *Military Telegraph*, 2:224; Halleck to Stager, 26 Feb. 1862; Stager to Halleck, 27 Feb. 1862; Halleck to Stager, 1 Mar. 1862, all in Caton Papers; Theodore Holt to Clowry, 18 Dec. 1864, Robert C. Clowry Papers, Manuscript Division, Library of Congress, Washington, D.C.

20. Rebecca Robbins Raines, *Getting the Message Through: A Branch History of the U.S. Army Signal Corps* (Washington, D.C.: Center of Military History, United States Army, 1996), 7–13; J. Willard Brown, *Signal Corps U.S.A. in the War of the Rebellion* (Boston: U.S. Veteran Signal Corps Association, 1896), 161–62.

21. Paul Joseph Schieps, "Albert James Myer, Founder of the Army Signal Corps: A Biographical Study" (Ph.D. diss., American University, 1966), 340–47; Brown, *The Signal Corps U.S.A.*, 48, 172–73.

22. For Rogers's background, see Henry J. Rogers Papers, Manuscript Division, Library of Congress, Washington, D.C.

23. Raines, *Getting the Message Through*, 17–23; Report of Capt. Samuel T. Cushing, Acting Chief Signal Officer, Army of Potomac, 23 May 1863, *Official Records*, 25.1:217–23; Maj. Gen. Daniel Butterfield to Maj. Gen. Couch, 30 Apr. 1863, and Butterfield to Maj. Gen. Joseph Hooker, 1 May 1863, *Official Records*, 25.2:304–5, 322–23; Capt. Joseph H. Spencer to Beardslee Magneto Electric Co., 24 Aug. 1863, 10 Sept. 1863, 15 Oct. 1863, 31 Oct. 1863; Maj. Nicodemus to William H. Coleman, 2 Mar. 1864; Capt. Henry S. Tafft to Capt. L. B. Norton, 31 Oct. 1863, all in Letters Sent, 1860–69 (Entry 1), Records of the Office of the Chief Signal Officer, RG 111, National Archives, Washington, D.C.

24. Charles F. Benjamin, "Recollections of Secretary Stanton," *Century*, March 1887, 761; Frank Abial Flower, *Edwin McMasters Stanton: The Autocrat of Rebellion, Emancipation, and Reconstruction* (New York: Saalfield, 1905), 216–22; Gideon Welles, *Diary of Gideon Welles* (Boston: Houghton Mifflin, 1911), 1:365; *Civil War Papers of George B. McClellan*, 70.

25. Brown, *Signal Corps U.S.A.*, 142–43, 180–81, 422–23; Plum, *Military Telegraph*, 2:103. For details of Myer's efforts to gain control of the field telegraphs and Stanton's dismissal of him, see Albert J. Myer Papers, Signal Corps Museum, microfilm held at Manuscript Division, Library of Congress, and Albert J. Myer Papers, Manuscript Division, Library of Congress, Washington, D.C.

26. Myer to Stager, 4 Nov. 1867, vol. 7, Letters Sent, 1860–69 (Entry 1), Records of the Office of the Chief Signal Officer, RG 111. Other correspondence between Myer and Stager in this entry details the cordial nature of their relationship.

27. Schieps, "Albert James Myer"; James Rodger Fleming, "Storms, Strikes, and Surveillance: The U.S. Army Signal Office, 1861–1891," *Historical Studies in the Physical and Biological Sciences* 30 (2000): 315–32.

28. Brown, *Signal Corps U.S.A.*, 168.

29. Entry of 5 Mar. 1863, bound vol. 11, Gross Papers; Plum, *Military Telegraph*, 1:326–28.

30. John Lonergan to William A. Simonds, 1 Aug. 1932, TAED X001E12A; George Bryan, *Edison: The Man and His Work* (Garden City, N.Y.: Garden City Publishing, 1926), 29.

31. Stuart McConnell, *Glorious Contentment: The Grand Army of the Republic, 1865–1900* (Chapel Hill: University of North Carolina Press, 1992).

32. Of the men discussed here, only Luther A. Rose's postwar activities are unknown. For Eckert and Stager, see their biographies in *American National Biography Online*, www.anb.org, accessed 26 July 2011. On Schnell, see an obituary in *Utica Saturday Globe*, 27 Sept. 1890, clipping in folder 3 of Joseph Schnell Papers, Manuscript Division, Library of Congress, Washington, D.C. On Van Duzer, see Van Duzer to Garfield, 20 Feb. 1877, James A. Garfield Papers, Manuscript Division, Library of Congress, Washington, D.C., and *Report of the Chief Signal Officer to the Secretary of War, for the Year 1868* (Washington, D.C.: GPO, 1868), 25–27.

33. [L. H. Smith], "Reasons Why," *Telegrapher* 1 (28 Nov. 1864): 22.

34. George Kennan to Hattie Kennan, undated Oct. 1862 and 18 Oct. 1863; Kennan to his family, 29 Nov. 1862, all in George Kennan Papers, Manuscript Division, Library of Congress, Washington, D.C.; entries of 11 Sept. and 12 Oct. 1862, "Official Records, Telegraph Office, Cairo, Ill., Aug. 15, 1862 to Dec. 22, 1862," Gross Papers; Clowry to R. H. Smith, 10 Apr. 1864; Clowry to Anson Stager, 20 June 1865, both in Clowry Papers; J. J. S. Wilson to Caton, 17 Oct. 1863, Caton Papers; Plum, *Military Telegraph*, 1:181, 244, 276.

35. By way of comparison, in the fall of 1864 "first-class" operators, those who could receive fifteen words a minute or more by sound, earned $110 to $120 a month on the commercial lines. In contrast, USMT pay ranged from $125 a month for assistant superintendents, $100 for chief operators, $60 to $100 for first-class operators, $50 for second-class operators (those unable to copy fifteen words a minute by ear), and $35 for unskilled laborers. Plum, *Military Telegraph*, 2:107; Directors' Minutebook, American Telegraph Company, WUTC; Abstract for Sept. 1863, Gross to his father, 4 Dec. 1863, Gross Papers; Clowry to unnamed recipient, 14 Feb. 1865, Clowry Papers.

36. Gross to Sam Bruch, 25 Jan. 1864; Gross to A. T. Langhorne, 5 Dec. 1864; Gross to Van Duzer, 17 Aug. 1865, Gross Papers.

37. On the NTU and incipient trade unionism among telegraphers, see Edwin Gabler, *The American Telegrapher: A Social History, 1860–1900* (New Brunswick: Rutgers University Press, 1988), 145–49. Material related to the abortive strike is in both Gross Papers and Plum, *Military Telegraph*, 2:113–16, 2:170–74.

38. Plum, *Military Telegraph*, 2:116–19; Andrews, "Southern Telegraph Company," 336–40; J. R. Dowell to William S. Morris, 1 June 1864, Morris Family Collection, Museum of the Confederacy, Richmond, Va.

39. Plum, *Military Telegraph*, 2:111–13; Clowry to Lt. Col. W. D. Green, 27 June 1864, Clowry Papers; entry of 3 December 1863, Rose Diary.

40. Sears, *Civil War Papers of McClellan*, 254; Plum, *Military Telegraph*, 1:145; undated 1862 newspaper clipping, Autograph Book, Schnell Papers; Dana, *Recollections of the Civil War*, 117; entries of 2, 5, and 22 June 1864, Rose Diary.

41. M. Adams and E. B. Smith to Gross, 18 Feb. 1864, Gross Papers; Clowry to Capt. George H. Smith, 16 Apr. 1864, Clowry Papers.

42. "Address of W. R. Plum, President of the Society of the United States Military Telegraph Corps," 19 August 1891, Gross Papers; "Re-Union Proceedings of the Society of the United States Military Telegraph Corps. Chicago, September 19–20, 1883"; Markle, *Telegraph Goes to War*, 8.

43. "Reunion Proceedings of the Society of the United States Military Telegraph Corps," 1885, 15; 1886, 6; 1890, 13.

44. *Report of the Committee of the Senate upon the Relations between Labor and Capital, and Testimony Taken by the Committee* (Washington, D.C.: GPO, 1885), 1:101–21, 1:185–97.

45. Gabler, *American Telegrapher*; "Eighteenth Reunion Proceedings of the Society of the United States Military Telegraph Corps," 3 and 16. For a firsthand account of working conditions of telegraphers and of the 1870 strike, see the diary of William Andrew Manning, a Cleveland telegrapher. William Andrew Manning Papers, Western Reserve Historical Society, Cleveland, Ohio.

46. Thompson, *Wiring a Continent*, chap. 20.

47. Andrews, "Southern Telegraph Company," 326–32; Eicher, *Longest Night*, 301.

48. "Memorandum of Agreement Entered into between S Bruch Capt & AQM Ast Supt US Military Telegraph and JJS Wilson Supt Caton Telegraph Company," 24 Sept. 1863, Caton Papers; "Sixth Annual Meeting of the North American Telegraph Association," 7–10 Oct. 1863, 6–9.

49. Norvin Green to Pinckney Green, 23 Oct. 1864 and 4 Feb. 1865, Green Family Papers, University of Kentucky Library, Lexington; American Telegraph Co. Directors Minutebook, WUTC; Illinois and Mississippi Telegraph Co. monthly statements from 1861 to 1865; S. Churchill to Caton, 31 March 1865, both in Caton Papers.

50. Thompson, *Wiring a Continent*, 398–401, 407–10; James D. Reid, *The Telegraph in America* (New York: Weed, Parsons, 1879), 484–85.

51. Plum, *Military Telegraph*, 2:346–48; John Van Duzer to Caton, 7 Apr. 1866; Account of U.S. Government with Illinois & Mississippi Telegraph Co. for June to Oct. 1861, 7 Nov. 1861, both in Caton Papers; entries of 30 June 1862, 5 Dec. 1865, and 8 Apr. 1866, Minutes of the Board of Directors of the Chicago and Mississippi Telegraph Company, WUTC.

52. Thompson, *Wiring a Continent*, 386; Bassnett to Caton, 15 Dec. 1861, Caton Papers; Kennan to his family, 2 Jan. and 7 Oct. 1863, 30 Mar. 1865, all in Kennan Papers.

53. Thompson, *Wiring a Continent*, 395; Green to Gross, 31 July 1865, 27 Oct. 1865, 13 Nov. 1865; Van Duzer to Gross, 27 Aug. 1865 and 5 Feb. 1866, all in Gross Papers.

54. Mark R. Wilson, *The Business of Civil War: Military Mobilization and the*

State, 1861–1865 (Baltimore: Johns Hopkins University Press, 2006), 1–4. Much of the stigma associated with Civil War procurement attached to Lincoln's first secretary of war, Simon Cameron, who was inept, corrupt, or, more generously, overwhelmed. Jean Baker, "Cameron, Simon," *American National Biography Online*, www.anb.org, accessed 1 Jan. 2011.

55. Bassnett to Caton, 15 Dec. 1861, Caton Papers; Thompson, *Wiring a Continent*, 396–98.

56. Thompson, *Wiring a Continent*, chap. 29. "Communication from the Secretary of State to Hon. Zachariah Chandler, Chairman of Committee on Commerce, Relative to Telegraphic Communication between the Eastern and Western Continents," 38th Cong. 1st sess., 9 June 1864, S. Misc. Doc. 123. For details of Western Union's work on the Russian Extension, consult the correspondence between Hiram Sibley and Perry McD. Collins in the Hiram Sibley Papers, Rush Rhees Library, University of Rochester, Rochester, N.Y.

57. Thompson, *Wiring a Continent*, 369–71.

58. David Hochfelder, "Taming the Lightning: American Telegraphy as a Revolutionary Technology, 1832–1860," (Ph.D. diss., Case Western Reserve University, 1999), 239–41.

59. Orton to George F. Davis, 8 Jan. 1866; Orton to J. W. Kirk, 13 Jan. 1866; Orton to Davis, 15 Jan. 1866, all in PLB WUTC; Thompson, *Wiring a Continent*, chap. 28.

60. S. Churchill to Caton, 22 and 24 Mar. 1865, Caton Papers. For details on Western Union's reaction to the startup of the United States Telegraph Company, see Thomas R. Walker to Hiram Sibley, 25 Mar. 1863; Emory Cobb to Sibley, 2 and 30 Dec. 1863, 3 and 19 Feb. 1864; O. H. Palmer to Sibley, 11 Jan. 1864; A. B. Smith to Sibley, 18 Apr. 1864, 12 and 16 May 1864; John Horner to Sibley, 25 December 1864, all in Sibley Papers.

61. The Caton Papers contain extensive correspondence on these negotiations between March and August 1866.

62. T. J. Stiles, *The First Tycoon: The Epic Life of Cornelius Vanderbilt* (New York: Alfred A. Knopf, 2009), 510–11.

Chapter 2. "As a Telegraph for the People It Is a Signal Failure":
The Postal Telegraph Movement

1. "Telegraphs for the United States," 25th Cong., 2d sess., 1837, H. Doc. 15. For a summary of postal operation of the telegraph, see Robert Luther Thompson, *Wiring a Continent: The History of the Telegraph Industry in the United States, 1832–1866* (Princeton: Princeton University Press, 1947), 20–34.

2. Morse strongly wished the Post Office to operate his telegraph, yet by 1841, after he had grown exasperated with congressional delays in appropriating funds, he started negotiations with a group of New York investors to fund a line between New York and Philadelphia. Morse to Alfred Vail, 16 Aug. 1841 and 13 Dec. 1841, Vail Telegraph Collection, Smithsonian Institution Archives, Washington, D.C.

3. Richard Bensel characterizes the federal government at 1860 as "a mere shell" with "only a token administrative presence." Bensel, *Yankee Leviathan: The Origins*

of Central State Authority in America, 1859–1877 (Cambridge: Cambridge University Press, 1990), ix. Morton Keller also refers to the "political malaise" of the 1850s that "had substantially reduced Congress's effectiveness as an instrument of government." Keller, *Affairs of State: Public Life in Late Nineteenth Century America* (Cambridge: Harvard University Press, 1977), 21. For a countervailing view, see William J. Novak, "The Myth of the 'Weak' American State," *American Historical Review* 113 (June 2008): 752–72, and an exchange on Novak's article in *American Historical Review* 115 (June 2010): 766–800.

4. The Morse patentees typically took half the stock in new telegraph ventures in this period. A typical line, like Jeptha Wade's Cleveland and Cincinnati Telegraph Company, was capitalized at $300 per mile; the Morse group received $150 per mile in stock. On the other hand, Morse offered to sell his patent to the federal government for $100,000—equal to the amount accruing from 667 miles of privately owned line.

5. For an extended discussion, see Bensel, *Yankee Leviathan*, especially chaps. 4 and 5, and Keller, *Affairs of State*.

6. The postal telegraph issue is perhaps the best-studied aspect of the history of the American telegraph industry. Daniel J. Czitrom, *Media and the American Mind: From Morse to McLuhan* (Chapel Hill: University of North Carolina Press, 1983), 3–29; Josh Wolff, " 'The Great Monopoly': Western Union and the American Telegraph, 1845–1893" (Ph.D. diss., Columbia University, 2008); Richard R. John, *Network Nation: Inventing American Telecommunications* (Cambridge: Harvard University Press, 2010), chaps. 3–5.

7. John Stuart Mill, *Principles of Political Economy, with Some of Their Applications to Social Philosophy* (New York: D. Appleton, 1893), 2:278; Richard T. Ely, *Problems of To-Day: A Discussion of Protective Tariffs, Taxation, and Monopolies* (New York: Thomas Y. Crowell, 1888), 251–73; Henry C. Adams, *Relation of the State to Industrial Action* (Baltimore: Guggenheimer, Weil, 1887). For a survey of the history of the concept of natural monopoly, see Manuela Mosca, "On the Origins of the Concept of Natural Monopoly: Economies of Scale and Competition," *European Journal of the History of Economic Thought* 15 (2008): 317–53. Libertarian economists contend that "natural" monopolies do not exist and that they are really creations of the state. See, for example, Thomas J. DiLorenzo, "The Myth of Natural Monopoly," *Review of Austrian Economics* 9 (1996): 43–58.

8. Ely, *Problems of To-Day*, 278.

9. Orton to Josiah King, 12 Mar. 1866, PLB WUTC; Edison to Wiman, 5 July 1884; Wiman to Edison, 7 July 1884, TAE Digital Edition, Docs. PA436 and D8471Q1.

10. "Letter from the Postmaster General," 39th Cong, 1st sess., S. Ex. Doc. No. 49, 1866, 11; WUTC Annual Report for 1878, 7; *A Talk on Telegraphic Topics. A Bid for Business* (New York: Francis Hart, [1881]), 36. (WorldCat and the finding aid for the Western Union Telegraph Company collection at the National Museum of American History both attribute this pamphlet to Erastus Wiman.) Forbes to William H. Haile, 19 Mar. 1883, quoted in Arthur Pier, *Forbes: Telephone Pioneer* (New York: Dodd, Mead, 1953), 143–45.

11. *Annual Report of the Atlantic & Pacific Telegraph Company* (New York: Van Kleeck, Clark, 1877, 5).

12. "Symbiosis," *Railway and Locomotive Engineering* 25 (July 1912): 270; John, *Network Nation*, 95–96, 164–67; James D. Reid, *The Telegraph in America* (New York: Weed, Parsons, 1879), 243–45, 477–81; *Talk on Telegraphic Topics*, 16; *Report of the Committee of the Senate upon the Relations between Labor and Capital, and Testimony Taken by the Committee* (Washington, D.C.: GPO, 1885), 1:1071 ; George Thurston to Andrew Carnegie, 6 Nov. 1867, Carnegie Steel Company Papers, Historical Society of Western Pennsylvania, Pittsburgh, Pa.; P&A Annual Report for 1871, 8–10; A&P Annual Report for 1875; Senate Committee on Interstate Commerce, To Amend the Communications Act of 1934: Hearings on S. 1335, 74th Cong., 1st sess., 1935, 67–75.

13. Ironically, Western Union had brought this suit to prevent the Pennsylvania Railroad from replacing its lines with those of the Postal company. The Pennsylvania had taken this step as part of a larger railroad war between its president Alexander J. Cassatt and Jay Gould's son George. Justice John Marshall Harlan dissented, interpreting the 1866 act as giving any telegraph company the "absolute right to put its wires and poles upon any post road,—a public highway established primarily for the public convenience,—if the ordinary travel on such road was not thereby interfered with." *Signal-Service and Telegraph Companies*, 42d Cong., 2d sess., 1872, H. Rep. 69, 36–37; *Western Union Telegraph Company v. Pennsylvania Railroad Company*, 195 U.S. 540 (1904); Senate Committee on Interstate Commerce, *To Amend the Communications Act of 1934, Hearings before a Subcommittee of the Committee on Interstate Commerce, Senate, on S. 1335*, 74th Cong., 1st sess., 1935; "Goulds Make Peace with Pennsylvania," *New York Times*, 3 Oct. 1907, 1; Edwin P. Hoyt, *The Goulds: A Social History* (New York: Weybright and Talley, 1969), 232–34; S. Walter Jones, *A Treatise on the Law of Telegraph and Telephone Companies* (Kansas City, Mo.: Vernon Law Book Company, 1906), 164–67.

14. S[imon]. N[ewton]. D[exter]. North, *History and Present Condition of the Newspaper and Periodical Press of the United States* (Washington, D.C.: GPO, 1884), 107; Frederic Hudson, *Journalism in the United States; From 1690 to 1872* (New York: Harper & Brothers, 1873), 617.

15. Michael J. Makley, *John Mackay: Silver King in the Gilded Age* (Reno: University of Nevada Press, 2009), 175–89; Robert G. Ingersoll, "Some Interrogation Points," *North American Review* 144 (March 1887): 222. Gould's quote is unattributed in Lewis Coe, *The Telegraph: A History of Morse's Invention and Its Predecessors in the United States* (Jefferson, N.C.: McFarland, 1993), 91.

16. *TAE Papers*, 2:789; *Relations between Labor and Capital*, 1:483; "Tests of the Synchronograph on the Telegraph Lines of the British Government," *Journal of the Franklin Institute* 145 (May 1898): 365–67; "Topics of the Times," *New York Times*, 5 May 1902, 8; "Topics of the Times," *New York Times*, 15 May 1902, 8; "Advances Made in Modern Telegraphy," *New York Times*, 25 May 1902, 28.

17. Reid, *Telegraph in America*, 587–590.

18. William Orton to Henry Weaver, 12 Sept. 1872, PLB WUTC; Reid, *Telegraph in America*, 735–36, 742; *TAE Papers*, 1:513–15, 531–33, 555–56; 2:340–44.

19. Stephenson's remarks are quoted in Edwin R. A. Seligman, *Railway Tariffs and the Interstate Commerce Law* (Boston: Ginn, 1887), 47.

20. Charles Albro Barker, *Henry George* (New York: Oxford University Press, 1955), 110–20.

21. John, *Network Nation*, 142–43, 193–94; Orton to S. A. B. Balcombe, 11 Nov. 1867; Orton to Stager, 26 Feb. 1869, both PLB WUTC; Henry Watterson, *"Marse Henry": An Autobiography* (New York: George H. Doran, 1919), 296; *Postal Telegraph. Speech of Hon. Charles A. Sumner, of California, in the House of Representatives, Saturday, March 8, 1884* (Washington, D.C.: n.p. 1884), 28; "Stock Jobbers at Work," *New York Times*, 7 Nov. 1884, 5.

22. Herbert L. Satterlee, *J. Pierpont Morgan: An Intimate Portrait* (New York: Macmillan, 1939), 206; Gustavus Myers, *History of the Great American Fortunes* (Chicago: Charles H. Kerr, 1908–11), 3:90.

23. Orton to Stager, 15 May 1868; Orton to J. S. Bedlow, 8 July 1868; Orton to Robert Clowry, 25 July 1868; Orton to Uriah H. Painter, 6 Jan. 1869; Orton to John Potter Stockton, 16 Mar. 1869, all in PLB WUTC; Thomas T. Eckert to James Garfield, 3 Aug. 1871; Garfield to Eckert, 8 Aug. 1871, both in James A. Garfield Papers, Manuscript Division, Library of Congress, Washington, D.C.; Frank Parsons, *The Telegraph Monopoly* (Philadelphia: C. F. Taylor, 1900), 93; "Lobbying with Telegraph Franks," *Chicago Daily Tribune*, 6 Feb. 1891, 9.

24. Statistical Notebooks, WUTC; and WUTC Annual Report for 1897.

25. House Committee on Post Offices and Post Roads, *Government Telegraph*, 53d Cong., 2d sess., 1894, 56; *Speech of Hon. Charles A. Sunmer, of California, in the House of Representatives, Saturday, February 28, 1885* (Washington: n.p., 1885), 15; *Speech of Hon. Charles A. Sumner . . . , March 8, 1884*, 28–29; Parsons, *Telegraph Monopoly*, 94–95.

26. Orton to Hiram Sibley, 7 Sept. 1869; Orton to Z. G. Simmons, 7 Sept. 1869, both in PLB WUTC.

27. "The Telegraph Lobby at Work," *New Orleans Times Picayune*, 23 Dec. 1868, 2; "Bribery and Corruption at Washington," *New York Herald*, 28 Feb. 1870; "The Western Union Lobby at Work," *New York Herald*, 1 Mar. 1870, 3; Orton to William E. Chandler, 13 Apr. 1868, quoted in Leon Burr Richardson, *William E. Chandler, Republican* (New York: Dodd, Mead, 1940), 83; Orton to Painter, 6 Jan. 1869, 16 Mar. 1869, 25 May 1870, all in PLB WUTC; Norvin Green to John G. Carlisle, 15 Dec. 1883, Norvin Green Papers, Filson Historical Society, Lexington, Ky. I thank Josh Wolff and Richard John for making their notes on the Green Papers available to me.

Painter was Clerk of the House Committee on Post Offices and Post Roads from 1865 to 1875 at a salary of $112 to $153 per month. I am grateful to Benjamin Hayes, Office of the Historian, U.S. House of Representatives, for this information.

28. "Hatch on Vanderbilt," 17 Oct. 1882, *Chicago Daily Tribune*, 12; "Proposing a Platform and a National Ticket," *New York Evening Telegram*, 9 Jan. 1884, 3; "A New Party Is Forming," *New York Times*, 5 Dec. 1890, 1; Senate Committee on Interstate Commerce, *The Anti-Monopoly League. Remarks of the Chairman, Mr. W. A. A. Carsey*, 50th Cong., 1st sess., 1888, 6; Morgan to Orton, 9 June 1866; Orton to Morgan, 18 July 1872, both in Edwin D. Morgan Papers, New York State Library, Albany; Green to Carsey, 28 Nov. 1890, PLB WUTC; Green to Robert M. Gallaway, 26 Mar. 1889; Green to John G. Moore, 25 Feb. 1888, both in Green Papers.

29. Gardiner G. Hubbard, "Government Control of the Telegraph," *North American Review* 325 (Dec. 1883): 522.

30. *Congressional Globe*, 39th Cong., 1st sess., 1866, 1029–31, 1518–22, and 2215–17; Canter Brown Jr., "The International Ocean Telegraph," *Florida Historical Quarterly* 68 (Oct. 1989): 135–59; Jay Cooke to Sherman, 20 Apr. and 22 June 1866, John Sherman Papers, Manuscript Division, Library of Congress, Washington, D.C.; Reid, *Telegraph in America*, 531.

31. House Committee on Appropriations, *Proceedings of the Committee on Appropriations in the Matter of the Postal Telegraph*, 42d Cong., 3d sess., H. Misc. Doc. 73, 25–26.

Yet many businessmen and newspaper editors complained of high rates and indifferent service. See John, *Network Nation*, 128–33, 146–47, 173–75; Richard A. Schwartzlose, *The Nation's Newsbrokers* (Evanston, Ill.: Northwestern University Press, 1990), 2:57–59, 79–80, 111–19; Makley, *John Mackay*, 174–79; meetings of 4 Feb. 1878, 1 Feb. 1881, 22 Mar. 1881, and 6 Nov. 1883, BoD CBOT; "Postal Telegraphy," *Chicago Daily Tribune*, 30 Jan. 1873, 2.

32. House Committee on Post Offices and Post Roads, *Postal Telegraph. Letter from the Postmaster General*, 40th Cong., 3d sess., 1869, H. Ex. Doc. 35, 18–22; Orton to Stager, 27 Nov. 1869, PLB WUTC.

33. Reid, *Telegraph in America*, 485; Orton to Z. G. Simmons, 8 Feb. 1869, PLB WUTC; "The Postal Telegraph," *Chicago Daily Tribune*, 18 Nov. 1882, 11; WUTC Annual Report for 1878, 9.

34. Maury Klein and Richard R. John give much higher figures for the amount of stock that Orton sequestered in the company's treasury, $11 million and $15 million, respectively. Yet Western Union's annual reports give this amount at $7.3 million, of which $1.3 million went to buy stock in the Atlantic and Pacific Telegraph Co. in 1879. Maury Klein, *The Life and Legend of Jay Gould* (Baltimore: Johns Hopkins University Press, 1986), 196; John, *Network Nation*, 167; "The Western Union Melon," *New York Times*, 2 Nov. 1878, 8; "Western Union Telegraph," *New York Times*, 9 Oct. 1879; WUTC Annual Reports for 1877, 1879, 1881, 1882, and 1885.

Green's views about how to manage the company were more in tune with those of prominent shareholders than Orton's had been. Orton was a salaried manager who owned only about one hundred shares of Western Union stock. Green, on the other hand, speculated avidly in stocks. At his death in early 1893, he left an estate worth about $2 million, mainly in Western Union stock. "Mr. Orton's Salary and Life-Insurance," *Chicago Daily Tribune*, 29 Apr. 1878, 8; "The Orton Estate," *Chicago Daily Tribune*, 30 Apr. 1878, 7; "Dr. Norvin Green's Death," *New York Times*, 13 Feb. 1893, 1. Green's personal letterbooks in the Green Papers contain much correspondence on his stock speculations.

35. Green to William H. Vanderbilt, 27 Aug. 1878, Green Papers; *Relations between Labor and Capital*, 1:878–81; Shelby M. Cullom, "The Government and the Telegraph," *Forum* 4 (Feb. 1888), 571.

36. Senate Committee on Post Offices and Post Roads, *Report on Postal Telegraph*, 48th Cong., 1st sess., 1884, S. Report 577, 1 and 5.

37. House Committee on Post Offices and Post Roads, *Union of the Telegraph and Postal System*, 40th Cong., 2d sess., 18 May 1868, H. Misc. Doc. 129, 4; Gardiner G. Hubbard, "The Proposed Changes in the Telegraphic System," *North American Review* 117 (July 1873): 101. See also Hubbard, "Government Control of the Telegraph," 531–32.

38. William Stanley Jevons, *Methods of Social Reform and Other Papers* (London: Macmillan, 1883), 293.

39. For favorable views of the British postal telegraph, see, for example, *Union of the Telegraph and Postal System*; *Postal Telegraph in the United States*, 41st Cong., 2d sess., 5 July 1870, H. Report 114; *Report of the Postmaster General* (18 Nov. 1871), 42d Cong., 2d sess., Ex. Doc. 1, Part 4, xxv–xxx; Parsons, *Telegraph Monopoly*, chap. 10.

40. Letter of transmittal from Postmaster General A. W. Randall, covering "Postal Telegraph," 40th Cong., 3d Sess., H. Ex. Doc. 35, 11 Jan., 1869, 1–8.

41. For an extended discussion on this point, see David Hochfelder, "A Comparison of the Postal Telegraph Movements in Great Britain and the United States, 1866–1900," *Enterprise and Society* 1 (2000): 739–61.

42. Henry C. Fischer and William H. Preece, "Joint Report upon the American Telegraph System," 183–86 and 196–200, British Post Office Archives, London. On the finances of the British postal telegraph, see Jeffrey Kieve, *The Electric Telegraph: A Social and Economic History* (Newton Abbot: David & Charles, 1973), chap. 9.

43. Parsons, *Telegraph Monopoly*, 158.

44. During the 1860s, Preece was a leading telegraph engineer for one of the British railways, and he staunchly opposed nationalization. Afterward, however, he was one of its warmest supporters. See Kieve, *Electric Telegraph*, 139. Preece's interview of Sunday, 20 Aug. 1893, with the *New York Sun* is reproduced in Edward Cecil Baker, *Sir William Preece, F.R.S.: Victorian Engineer Extraordinary* (London: Century Benham, 1976), 244–45.

45. *Congressional Globe*, 39th Cong., 1st sess., 21 Dec. 1865 and 23 Feb. 1866, 115, 979–80.

46. *Telegraph in Connexion with the Postal System*, 39th Cong., 1st sess., 2 June 1866, S. Ex. Doc. 49.

47. *Congressional Globe*, 39th Cong., 1st sess., 26 Feb., 20 Mar., and 27 Apr. 1866, 1029–31, 1518–22, and 2215–17; Canter Brown Jr., "The International Ocean Telegraph," *Florida Historical Quarterly* 68 (October 1989): 135–59.

48. For a list of the incorporators of the National Telegraph Company, see "Washington News," *New York Times*, 6 Apr. 1866, 1. For Nye's remarks, see *Congressional Globe*, 39th Cong., 1st sess., 29 June 1866, 3485. For details on Sherman's work on this bill, see William D. Snow to Sherman, 7 and 15 Apr. 1866; Warner M. Bateman to Sherman, 10 Apr. 1866, George B. Senter to Sherman, 17 June 1866; Jay Cooke to Sherman, 1 Aug. 1866, all in Sherman Papers; Thomas Ewing to John Sherman, 17 Nov. 1866, Ewing Family Papers, Manuscript Division, Library of Congress, Washington, D.C.

49. For a somewhat parallel discussion of the National Telegraph Act, see John, *Network Nation*, 116–24. For details on the legislative history of the Act and Western Union's reaction to it, see Wolff, "'The Great Monopoly'," 195–212.

50. Norvin Green to George Douglass, 5 July 1867, George Douglass Papers, Fil-

son Historical Society, Lexington, Ky. I am grateful to Richard R. John for making his notes available to me. House Committee on Appropriations, *Signal-Service and Telegraph Companies*, 42d Cong., 2d sess., 9 May 1872, H. Report 69, 36–37.

51. William Orton, *Government Telegraphs* (New York: Russell's American Steam Printing House, 1870), 28.

52. James Rodger Fleming, *Meteorology in America, 1800–1870* (Baltimore: Johns Hopkins University Press, 1990), 141–62; Rebecca Robbins Raines, *Getting the Message Through: A Branch History of the U.S. Army Signal Corps* (Washington, D.C.: Center of Military History, United States Army, 1996), 45–69; Erik D. Craft, "The Value of Weather Information Services for Nineteenth-Century Great Lakes Shipping," *American Economic Review* 88 (Dec. 1998): 1059–76.

53. From 1860 to 1869, the army communications branch performed purely military duties and was known as the Signal Corps. However, after 1870 the army referred to its weather reporting system as the Signal Service and its purely military communications arm as the Signal Corps. After 1891, when the Department of Agriculture took over the weather reporting service, the army again referred to its entire communications branch as the Signal Corps.

The Signal Corps began constructing telegraph lines for military and other purposes in 1873, especially to connect forts in the West and lighthouses and life-saving stations on the Atlantic coast. The Signal Corps had a policy of abandoning these lines as soon as commercial telegraph service reached those points. Raines, *Getting the Message Through*, 50–52.

54. *Report of the Chief Signal Officer to the Secretary of War, for the Year 1870* (Washington, D.C.: GPO, 1870), 8–10; Orton to C. S. Cutler, 18 Mar. 1871; Orton to Thomas Eckert, 7 Nov. 1871, both in PLB WUTC.

55. For specifics on the relationship between the Signal Service and all telegraph companies, see summaries of correspondence in entries 6 and 8, Records of the Weather Bureau, RG 27, National Archives, Washington, D.C.; CSO Annual Report for 1872, 88–89.

56. CSO Annual Report for 1874, 380–83; Orton to Morgan, 5 Sept. 1877, Morgan Papers.

57. *Postal Telegraph. Speech of Hon. Frank W. Palmer, of Iowa, in the House of Representatives, February 17, 1872* (Washington, D.C.: Congressional Globe Office, 1872), 3; Harry James Brown and Frederick D. Williams, eds., *The Diary of James A. Garfield* (East Lansing: Michigan State University Press, 1967–81), entries for 29 Mar. and 23 Apr. 1872, 2:36–37 and 45; Gardiner Hubbard to Gertrude Hubbard, 7 May 1872, Hubbard Family Papers, Manuscript Division, Library of Congress, Washington, D.C.

The Signal Service permitted a few New Jersey shore hotels to connect to its network and handled some press traffic for western newspapers in remote towns. See, for example, Myer to W. B. Crooks, 13 Aug. 1877; W. B. Hazen to Martin Maginnis, 11 Mar. 1881, both in Letters Sent, Chief Signal Officer, entry 12, RG 27. On Orton's opposition to the construction of Signal Service lines, see Orton to Morgan, 5 Sept. 1877, Morgan Papers.

58. For correspondence between the Signal Service and telegraph company of-

ficials on this issue, see entries 21, 22, and 23, RG 27. For an overview, see the CSO Annual Report for 1874, 103–4, 380–83. On Gould's interest in regaining this business, see Gould to Chandler, 6 and 7 Dec. 1874, William E. Chandler Papers, Manuscript Division, Library of Congress. I am grateful to Richard R. John for making his notes available to me. See also Orton to Henry Weaver, 16 July 1875, PLB WUTC; Reid, *Telegraph in America*, 451–54; A&P Annual Report for 1875, WUTC; Henry W. Howgate to Myer, 13 May and 17 July 1877, Albert J. Myer Papers, Signal Corps Museum, microfilm held at Manuscript Division, Library of Congress, Washington, D.C.

59. See, for example, American Union Telegraph Co. to Signal Service, 20 Jan. and 17 May 1880, entry 8, RG 27.

60. Gaillard Hunt, *Israel, Elihu and Cadwallader Washburn: A Chapter in American Biography* (New York: Macmillan, 1925), 239, 368–70; *Postal Telegraph. Speech of Hon. C.C. Washburn, of Wisconsin, Delivered in the House of Representatives, December 22, 1869* (Washington, D.C.: F. & J. Rives & Geo. A. Bailey, 1869); *Postal Telegraph in the United States; Union of the Telegraph and Postal System;* John, *Network Nation*, 124–26.

61. *Postal Telegraph in the United States;* Senate Committee on Post Offices and Post Roads, *Postal Telegraph System*, 41st Cong., 2d sess., 31 Jan. 1870, H. Report 18; *Congressional Globe*, 41st Cong., 2d sess., 27 Jan. and 5 July 1870, 824 and 5178; 42d Cong., 2d sess., 5 Dec. 1871, 27 Jan. 1872, 27 Apr. 1872, and 16 May 1872, 17–18, 635, 2830–32, 3554–62; 42nd Cong., 3d sess., 16 Dec. 1872, 207.

62. John, *Network Nation*, 135–36; Orton to Morgan, 20 Dec. 1867, Morgan Papers; *Chicago Daily Tribune*, 8 Feb. 1868, 2; "The Western Union Telegraph Job," *Woodhull & Claflin's Weekly*, 30 Dec. 1870; "Draft of proposition from the W.U. Tel. Co. to sell," 1 Dec. 1870, John A.J. Creswell Papers, Manuscript Division, Library of Congress, Washington, D.C.; *Congressional Globe*, 42d Cong., 2d sess., 5 Dec. 1871, 16–20; Orton to Stager, 24 Jan. 1871, PLB WUTC; "Speculation and Advance in Western Union Telegraph Co. Stock," *Telegrapher*, 4 Mar. 1871, 219; "The Future of the Telegraph in the United States," *Telegrapher* 8, 23 Sept. 1871, 36; Orton, *Government Telegraphs*, 45; Gardiner Hubbard to Gertrude Hubbard, 20, 27, and 29 Nov. and 2 and 16 Dec. 1870, Hubbard Papers; Brown and Williams, *Diary of James A. Garfield*, entry for 23 Dec. 1872, 2:128.

63. Richard R. John notes that the Butler Amendment was the culmination of Gould's attempts (working through Butler) since 1875 to erode the importance of Western Union's exclusive right-of-way contracts. John, *Network Nation*, 132, 164–66. For Western Union's reaction, see Green to William A. J. Sparks, 24 Mar. 1879; Green to William Vanderbilt, 2 Apr. 1879; and Green to Alonzo Cornell, 11 Apr. 1879, all in PLB WUTC. On the 1875 measure see *Congressional Record*, 43rd Cong., 2d sess., 17 Feb. 1875, 1419–24; House Judiciary Committee, "In the Matter of the Western Union Telegraph Co.," 43d Cong., 2d sess., 20–23 Feb. 1875, unpublished hearings. Letters written by Gould at the time showed that he was actively working behind the scenes through his ally Butler and lobbyists Gardiner Hubbard and William E. Chandler. Gould to Chandler, 6, 10, 15, and 27 Jan. and 21, 22, 26 Feb. 1875, Chandler Papers.

64. "The Telegraph Companies," *Manufacturer and Builder* 6 (Dec. 1874): 276; "Telegraphic Monopoly," *Manufacturer and Builder* 13 (Feb. 1881): 26.

65. "Talk about a Postal Telegraph," *New York Times,* 17 Jan. 1881, 1; "Telegraph," *Chicago Daily Tribune,* 18 Jan. 1881, 2; *Proceedings of the Eleventh Annual Meeting of the National Board of Trade, Held in Washington, December, 1880* (Boston: Tolman & White, 1881), 153–59; *Proceedings of the Twelfth Annual Meeting of the National Board of Trade, Held in Washington, January, 1882* (Boston: Tolman & White, 1882), 90–111 (I am grateful to Richard R. John for making copies available to me); meetings of 1 and 22 Mar. 1881, BoD, CBOT.

66. American Chamber of Commerce to Chicago Board of Trade, 29 Aug. 1883, BoD, CBOT. Abbott's editorial in the *Christian Union,* 30 Aug. 1883; John Wanamaker, *An Argument in Support of the Limited Post and Telegraph by the Postmaster-General* (Washington, D.C.: GPO, 1890), 182–223; Stuart B. Kaufman, ed., *The Samuel Gompers Papers* (Urbana: University of Illinois Press, 1986), 1:287–88; *Relations between Labor and Capital,* 1:891. Good accounts of the strike are Edwin Gabler, *The American Telegrapher: A Social History, 1860–1900* (New Brunswick: Rutgers University Press, 1988), 3–29; Vidkunn Ulriksson, *The Telegraphers: Their Craft and Their Unions* (Washington, D.C.: Public Affairs Press, 1953), 32–49. Gabler quotes Green's boast on 27, Ulriksson on 49.

67. Nathaniel P. Hill, *Postal Telegraph. Speech of Hon. N. P. Hill, of Colorado, Delivered in the Senate of the United States, January. 14, 1884* (Washington, D.C.: n.p., 1884); Green to Orville Platt, 25 Jan. 1884; Green to John Hay, 21 Feb. 1884; Green to Alonzo B. Cornell, 3 Apr. 1884; Green to James Beck, 13 June 1884, all in PLB WUTC.

68. John, *Network Nation,* 172; Makley, *John Mackay,* 159–89; Reid, *Telegraph in America,* 2d ed. (1886), 755–88; Senate Committee on Post Offices and Post Roads, *Statements of Robert Garrett and Other Officers of the Baltimore and Ohio Telegraph Company,* 48th Cong., 1st sess., 9 Feb. 1884.

69. Wanamaker, *An Argument in Support of the Limited Post and Telegraph;* Walter Clark, "The Telegraph and the Telephone Properly Parts of the Post Office System," *Arena* 5 (Mar. 1892): 471; William L. Wilson, "Why I Oppose Governmental Control of the Telegraph," *Arena* 15 (Jan. 1896): 247; "The Western Union vs. The Pennsylvania," *Outlook* 73 (31 Jan. 1903): 235–36.

70. "National Ownership of the Telegraph," *Independent* 63 (22 Aug. 1907): 460–61.

71. On the air of inevitability surrounding government control of the wires, see "Washington Notes," *Journal of Political Economy* 22 (Apr. 1914): 392–94; Jeremiah W. Jenks, "The Trend toward Government Management of Business," in *Shall the Government Own and Operate the Railroads, the Telegraph and Telephone Systems? Shall the Municipalities Own Their Utilities? The Affirmative Side* (New York: National Civic Federation, 1915), 5–18; *Final Report of the Commission on Industrial Relations* (Washington, D.C.: GPO, 1915), 104–8.

72. WWP 27:260, 50:249–50; *Government Ownership of Electrical Means of Communication,* 63d Cong., 2d sess., 1914, S. Doc. 399; John, *Network Nation,* 366.

73. E. David Cronon, ed., *The Cabinet Diaries of Josephus Daniels, 1913–1921* (Lincoln: University of Nebraska Press, 1963), 123, 137, 316;. Josephus Daniels to Burleson, 15 Dec. 1917; Burleson to Thomas R. Marshall, 4 Mar. 1918, both in Albert S. Burleson

Papers, Manuscript Division, Library of Congress; House Committee on Interstate and Foreign Commerce, *Federal Control of Systems of Communication: Hearings before the Committee on Interstate and Foreign Commerce of the House of Representatives*, 65th Cong., 2d sess., 2 July 1918. For an excellent summary, see Christopher N. May, *In the Name of War: Judicial Review and the War Powers since 1918* (Cambridge: Harvard University Press, 1989), chap. 2.

74. David Lewis to Wilson, 15 July 1918, WWP 48:623; "When Washington Holds the Wires," *Literary Digest*, 20 July 1918.

75. Postal Telegraph Company officials refused to provide accurate financial information and balked at handling government messages because they paid less than commercial messages. Mackay was a fierce critic of postal control and cooperated with Burleson as little as he could get away with. See material in Records Relating to Operation of the Telephone, Telegraph, and Cable Companies by the Postmaster General (entry 38), Records of the Post Office Department, RG 28, National Archives, Washington, D.C.

76. *Federal Control of Systems of Communication*, 48–49; "Mr. Burleson on the Grill," *Literary Digest*, 19 Apr. 1919, 14–15; Joseph Tumulty to Wilson, 8 May 1919, WWP 58:560–61. Wartime control of the wires caused Richard T. Ely, a longtime advocate of public ownership, to reverse his position. Benjamin G. Rader, *The Academic Mind and Reform: The Influence of Richard T. Ely in American Life* (Lexington: University of Kentucky Press, 1966), 224–25.

77. Parsons, *Telegraph Monopoly*, 11–21.

78. Figures about congressional consideration of telegraph reform measures are from Menahem Blondheim, "Rehearsal for Media Regulation: Congress versus the Telegraph-News Monopoly, 1866–1900," *Federal Communications Law Journal* 56 (Mar. 2004): 306.

79. Historians have become increasingly skeptical of this division. See, for example, the exchange between Richard Schneirov, Rebecca Edwards, and James L. Huston in the July 2006 issue of the *Journal of the Gilded Age and Progressive Era* and the forum "Should We Abolish the 'Gilded Age'?" in the October 2009 issue of the same journal. Regarding the role of federal regulatory power, Williamjames Hull Hoffer argues for the existence of a "second, intermediate state" in this period between a Jacksonian "first state" and a Progressive–New Deal "third state." Hoffer, *To Enlarge the Machinery of Government: Congressional Debates and the Growth of the American State, 1858–1891* (Baltimore: Johns Hopkins University Press, 2007), ix–xii.

80. Henry George Jr. *The Life of Henry George* (New York: Doubleday and McClure, 1900; repr., New York: Chelsea House, 1981), 183–86; Edward Bellamy, "First Steps toward Nationalism," *Forum* 10 (Oct. 1890): 174–76; Ely, *Problems of To-Day*; Edwin R.A. Seligman, "Government Ownership of Quasi-Public Corporations," *Gunton's Magazine* 20 (Apr. 1901): 305–22; Seligman, *Railway Tariffs and the Interstate Commerce Law*, 2, 45–47, 64; Simon Newcomb, "The Problem of Economic Education," *Quarterly Journal of Economics* 7 (July 1893): 375–99; "A Conversation with Professor Frank Parsons," *Arena* 27 and 28 (Feb., Mar., and Aug. 1902): 179–87, 297–302, 188–92; Parsons, *Telegraph Monopoly*; Frank Parsons, *The City for the People* (Phila-

delphia: C. F. Taylor, 1901); Rader, *Academic Mind and Reform*, 87–95; Arthur Mann, *Yankee Reformers in the Urban Age* (Cambridge: Harvard University Press, 1954).

81. John, *Network Nation*; Mark Lloyd, *Prologue to a Farce: Communication and Democracy in America* (Urbana: University of Illinois Press, 2006), chap. 3.

Chapter 3. "There Is a Public Voracity for Telegraphic News":
The Telegraph, Written Language, and Journalism

1. "Influence of the Telegraph upon Literature," *Democratic Review* 22 (May 1848): 409–13. Geoffrey Nunberg attributes the article to Conrad Swackhamer; http://peo ple.ischool.berkeley.edu/~nunberg/texting.html, accessed 17 May 2011.

2. Jeffrey Meyers, *Hemingway: A Biography* (New York: Harper & Row, 1985), 94.

3. James Carey argues that "the telegraph reworked the nature of written language and finally the nature of awareness itself," though he concedes that "much additional work needs to be done on the effects of the telegraph on language." James Carey, *Communication as Culture: Essays on Media and Society* (Boston: Unwin Hyman, 1989), 210–11.

4. Statistical Notebooks, WUTC. In 1861 the NYAP moved its headquarters into the American Telegraph Company building at 145 Broadway; after Western Union bought out the American in 1866, it became Western Union's headquarters. The WAP obtained office space at 145 Broadway over the objection of NYAP in 1871. Richard A. Schwartzlose, *The Nation's Newsbrokers* (Evanston, Ill.: Northwestern University Press, 1990), 1:223, 2:73.

5. Morse based his code on the quantities of type found in a printing office. See Edward Lind Morse, ed., *Samuel F. B. Morse: His Letters and Journals* (Boston: Houghton Mifflin, 1914), 2:61–69, and Morse to Charles T. Bull, 20 Dec. 1845, Samuel F. B. Morse Papers, Manuscript Division, Library of Congress, Washington, D.C.

Little evidence exists to support later claims that Alfred Vail invented Morse Code. To the contrary, Vail himself credited Morse with devising the alphabet code bearing his name. See David Hochfelder, "Two Controversies in the Early History of the Tele-graph," *IEEE Communications Magazine* 48 (Feb. 2010): 28–32; Alfred Vail to Messrs S. Vail & Son, 7 Feb. 1838, Vail Telegraph Collection; Alfred Vail, *The American Elec-tro Magnetic Telegraph: With the Reports of Congress, and a Description of all Telegraphs Known, Employing Electricity or Galvanism* (Philadelphia: Lea & Blanchard, 1845), 30, 168.

6. Morse to Vail, 29 May 1844; John Wills to Vail, 28 Mar. 1848; Vail to Faxton, 8 Oct. 1845, Vail Telegraph Collection, Smithsonian Institution Archives, Washington, D.C.; Cyrus Mendenhall to Charles Mendenhall, 4 Nov. 1872, Mendenhall Family Pa-pers, Folio, Cincinnati Historical Society, Cincinnati, Ohio. Morse gave Mendenhall the forty-six-inch-long message tape to keep as a souvenir.

7. For the use of abbreviations and codes as an important part of operators' work culture, see W. J. Johnston, *Lightning Flashes and Electric Dashes* (New York: W. J. Johnston, 1877); Walter Polk Phillips, *Sketches Old and New* (New York: George Mun-ro's Sons, 1897); Edwin Gabler, *The American Telegrapher: A Social History, 1860–1900* (New Brunswick: Rutgers University Press, 1988), 79–85. Psychologists William Lowe

Bryan and Noble Harter discovered that learning to receive Morse Code by sound was similar to learning a foreign language. Bryan and Harter, "Studies in the Physiology and Psychology of the Telegraphic Language," *Psychological Review* 4 (Jan. 1897): 27–53; and "Studies on the Telegraphic Language. The Acquisition of a Hierarchy of Habits," *Psychological Review* 6 (July 1899): 345–75.

8. Engineer's Reports, 27 Aug. 1863, 27 Jan. 1864, and 26 Jan. 1865, American Telegraph Co. File, WUTC.

9. WUTC Annual Report for 1888, 10; WUTC Annual Report for 1889, 9.

10. Louisa May Alcott, *Little Women; or, Meg, Jo, Beth, and Amy* (Boston: Roberts Bros., 1869), chap. 15, "A Telegram."

11. Richard B. DuBoff, "Business Demand and the Development of the Telegraph in the United States, 1844–1860," *Business History Review* 54 (Winter 1980): 470; Alexander Jones, *Historical Sketch of the Electric Telegraph, including Its Rise and Progress in the United States* (New York: George P. Putnam, 1852), 110; House Committee on Appropriations, *Proceedings of the Committee on Appropriations in the Matter of the Postal Telegraph*, 42d Cong., 3d sess., 1873, H. Misc. Doc. 73, 25–26; Green to Vilas, 17 Nov. 1887, PLB WUTC.

12. At their 1882 peak, night messages accounted for 13 percent of total message traffic, dropping to a low of 6.5 percent in 1899. Statistical Notebooks, WUTC; Western Union Annual Reports for 1910 and 1912. On Night Letters, see Richard R. John, *Network Nation: Inventing American Telecommunications* (Cambridge: Harvard University Press, 2010), 349–51.

13. For a technical description of the multiplex printer, see M. D. Fagen, ed., *A History of Engineering and Science in the Bell System: The Early Years, 1875–1925* (New York: Bell Telephone Laboratories, 1975), 755–58.

14. "Western Union," *Fortune*, Nov. 1935, 90–96; "Onward & Upward with the Arts," *New Yorker*, 21 Dec. 1940, 35. For a representative advertisement, see *Life*, 23 Dec. 1946, 4.

15. George B. Prescott, *History, Theory, and Practice of the Electric Telegraph* (Boston: Ticknor and Fields, 1860), 339, 349–50; Edwin Herbert Lewis, *Business English* (Chicago: La Salle Extension University, 1915), appendix p. 10; Roy Davis and Clarence H. Lingham, *Business English and Correspondence* (Boston: Ginn, 1914), 283–84; George Ade, *True Bills* (New York: Harper and Brothers, 1905), 150.

16. M. Bare & Co. Records, box 1, folder 3, Cincinnati Historical Society.

17. Halleck to Stager, 26 Feb. 1862; Stager to Halleck, 27 Feb. 1862; Halleck to Stager, 1 Mar. 1862, all in John Dean Caton Papers, Manuscript Division, Library of Congress, Washington, D.C.; Gross to Stager, 13 Oct. 1864, William L. Gross Papers, Western Reserve Historical Society, Cleveland, Ohio; Theodore Holt to Clowry, 18 Dec. 1864, Robert C. Clowry Papers, Manuscript Division, Library of Congress, Washington, D.C. William R. Plum calculated that commanders in Missouri, Arkansas, and Kansas sent a total of about 650 telegrams per day in 1864. Plum, *The Military Telegraph during the Civil War in the United States* (Chicago: Jansen, McClurg, 1882), 2:224.

18. David Hochfelder, "A Comparison of the Postal Telegraph Movements in Great Britain and the United States, 1866–1900," *Enterprise and Society* 1 (2000):

739–61; William Stanley Jevons, *Methods of Social Reform and Other Papers* (London: Macmillan, 1883), 301; Henry James, *In the Cage* (London: Hesperus Press, 2002), 18.

19. Memorandum on Day and Night Letters, 4 Feb. 1915, box 10, AT&T Archives, Warren, N.J.

20. For a technical discussion of inductive loading, see Fagen, *History of Engineering and Science in the Bell System*, 812–14, 977–87.

21. Sereno S. Pratt, *The Work of Wall Street* (New York: D. Appleton, 1903), 140–42; *Telegraph Age*, 16 Feb. 1906, 69; Gold and Stock Telegraph Co. President's Annual Report, 24 Sept. 1877, WUTC.

22. "How Is the Atlantic Telegraph to Affect the Condition of the World?" *New York Herald*, 9 Aug. 1858; Frederic Hudson, *Journalism in the United States, from 1690 to 1872* (New York: Harper & Brothers, 1873), 606. Dollar conversion is from www .measuringworth.com, accessed 30 June 2011.

23. George W. Smalley, *Anglo-American Memories* (London: Duckworth, 1910), 145; "Intellectual Gymnastics," *Telegrapher* 7, 10 Sept. 1870, 20; William Shirer, *20th Century Journey: A Memoir of a Life and the Times* (New York: Simon and Schuster, 1976), 1:283–84.

24. For general histories of wire-service journalism, see Hudson, *Journalism in the United States*, chaps. 37 and 38; Victor Rosewater, *History of Cooperative News-Gathering in the United States* (New York: D. Appleton, 1930); Schwartzlose, *Nation's Newsbrokers*; Menahem Blondheim, *News over the Wires: The Telegraph and the Flow of Public Information in America, 1844–1897* (Cambridge: Harvard University Press, 1994).

25. James L. Crouthamel, *Bennett's* New York Herald *and the Rise of the Popular Press* (Syracuse, N.Y.: Syracuse University Press, 1989), 43–47; WUTC Annual Report for 1869, 27–28. The formation of the NYAP formalized arrangements for telegraphic newsgathering that had been in place since 1846. See Schwartzlose, *Nation's Newsbrokers*, 1:89–121 and Blondheim, *News over the Wires*, chap. 3 for a full discussion.

26. Jones, *Historical Sketch of the Electric Telegraph*, 128–32; *New Hampshire Patriot & State Gazette*, 10 Feb. 1848, 2; 12 Feb. 1848, 3.

27. During the Prince of Wales's tour of the United States and Canada in 1860, a correspondent for the *New York Herald* reportedly blocked rival correspondents from using the telegraph by having the operator at Niagara Falls send Books of the Bible until his own dispatch was ready. "Telegraphing the Scriptures," *San Francisco Daily Evening Bulletin*, 7 Jan. 1861, 4; Thomas Low Nichols, *Forty Years of American Life* (London: J. Maxwell, 1864), 1:327–28.

28. For good summaries, see Blondheim, *News over the Wires*, 163–68; Schwartzlose, *Nation's Newsbrokers*, 2:91–104.

29. For information on press association wire leases, see Schwartzlose, *Nation's Newsbrokers*, 2:111–17, 252; "Effects on Employment of the Printer Telegraph for Handling News," *Monthly Labor Review* 34 (Apr. 1932): 753–58.

30. Morse to Vail, 25 May 1844, Vail Telegraph Collection. Craig's circular is quoted in Schwartzlose, *Nation's Newsbrokers*, 1:180–81.

31. Menahem Blondheim, "'Public Sentiment Is Everything': The Union's Public Communications Strategy and the Bogus Proclamation of 1864," *Journal of Ameri-*

can History (2002): 869–99; Richard B. Kielbowicz, "The Telegraph, Censorship, and Politics at the Outset of the Civil War," *Civil War History* 15 (1994): 95–118.

32. L[awrence]. A. Gobright, *Recollection of Men and Things at Washington, during the Third of a Century* (Philadelphia: Claxton, Remsen & Haffelfinger, 1869), 317; "How the News Was Received in Brooklyn," *New York Times*, 23 July 1861, 8.

33. *Allegations of Government Censorship of Telegraphic News Reports*, unpublished hearings of House Committee on the Judiciary, 37th Cong., 2d sess., 1862, HJ-T.1–12 (consulted on microfilm at the New York State Library, Albany). For examples of suppressed dispatches, see Thayer to F. W. Seward, 26 Oct. 1861; A. B. Talcott to F. W. Seward, undated Oct. 1861; Thayer to F. W. Seward, 2 Nov. 1861, all in William H. Seward Papers, Manuscript Division, Library of Congress, Washington, D.C.

34. Stager to H. H. Ward, 1 July 1866, Seward Papers.

35. Testimony of Samuel Wilkeson, 24 Jan. 1862; testimony of H. E. Thayer, 24 Jan. 1862; testimony of Lawrence Gobright, 5 Feb. 1862, all in *Allegations of Government Censorship*. Gobright, *Recollections of Men and Things*, 317–24.

36. Gobright to Charles Fulton and Daniel Craig, 29 and 30 June 1862, Telegrams Sent, RG 107.

37. David Homer Bates Diary, entries for 20 and 21 May 1864, Manuscript Division, Library of Congress, Washington, D.C.; "The Forged Proclamation," *New York Times*, 19 May 1864, 8; "The Forged Proclamation—Arrest of the Culprit," *New York Times*, 21 May 1864, 4; "The Bogus Call for Four Hundred Thousand Men," *New York Herald*, 19 May 1864, 4.

38. Donald Lewis Shaw, "At the Crossroads: Change and Continuity in American Press News, 1820–1860," *Journalism History* 8 (Summer 1981): 38–50; Donald Lewis Shaw, "Bias in the News: A Study of National Presidential Campaign Coverage in the Wisconsin English Daily Press, 1852–1916" (Ph.D. diss., University of Wisconsin, 1966); Donald Lewis Shaw, "News Bias and the Telegraph: A Study of Historical Change," *Journalism Quarterly* 44 (Spring 1967): 3–12. By 1888 Western Union leased some twelve thousand miles of wire to press associations, allowing them to use their own operators and to send news items at times of their own choosing.

David Mindich gives a useful distinction between the inverted-pyramid story and the ideal of objectivity. He identifies five attributes of journalistic objectivity: the inverted-pyramid structure, reliance on verifiable facts, detachment, nonpartisanship, and balance. However, Mindich and others disagree on the timing of the rise of objectivity as a core value of journalism. Dan Schiller finds the ideal of objectivity present even before the commercial introduction of the telegraph. Mindich precisely dates the origin of the inverted-pyramid story to Edwin Stanton's official dispatch announcing President's Lincoln's death on 15 April 1865. Richard Schwartzlose takes Craig's 1854 circular as "journalism's first expression of the principle of objectivity." However, Michael Schudson and Richard Kaplan attribute the widespread acceptance of the ideal of journalistic objectivity to the shedding of newspapers' partisan identities in the early twentieth century. Dan Schiller, *Objectivity and the News: The Public and the Rise of Commercial Journalism* (Philadelphia: University of Pennsylvania Press, 1981), 1–11; David Mindich, *Just the Facts: How "Objectivity" Came to De-*

fine American Journalism (New York: New York University Press, 1998), 8–10, 64–75; Schwartzlose, *Nation's Newsbrokers*, 1:154, 180–81; Michael Schudson, *Discovering the News: A Social History of American Newspapers* (New York: Basic Books, 1978), 3–8; Richard L. Kaplan, *Politics and the American Press: The Rise of Objectivity, 1865–1920* (New York: Cambridge University Press, 2002), 140–41.

39. "Bulletins of the Battle," "Effects of the Battle in Washington," and "Special Dispatches from Washington," all in *New York Times*, 22 July 1861, 1; "How the News Was Received in Brooklyn," *New York Times*, 23 July 1861, 8; Mary Boykin Chesnut, *A Diary from Dixie* (New York: D. Appleton, 1906), 171.

40. Chesnut, *Diary from Dixie*, 177. Shaw later led the black Fifty-fourth Massachusetts Regiment, fictionalized in the film *Glory*. Oliver Wendell Holmes, "Bread and the Newspaper," *Atlantic Monthly* 8 (Sept. 1861): 346–52. The Curtis quotation is from Charles Royster, *The Destructive War: William Tecumseh Sherman, Stonewall Jackson, and the Americans* (New York: Alfred A. Knopf, 1991), 246–47. An image of Prang's War Telegram Marking Map is available through the State Library of Massachusetts at http://archives.lib.state.ma.us/handle/2452/50832, accessed 25 Mar. 2012.

41. Signal Service telegrams from 1877 Great Railroad Strike, Hayes Presidential Library, Fremont, Ohio. See also Robert V. Bruce, *1877: Year of Violence* (Chicago: Ivan R. Dee, 1987 [1959]), 209–13; James Rodger Fleming, "Storms, Strikes, and Surveillance: The U.S. Army Signal Office, 1861–1891," *Historical Studies in the Physical and Biological Sciences* 30 (2000): 315–32.

42. "By Telegraph," *Macon Telegraph and Messenger*, 20 July 1877, 1; "In a Ferment," *Chicago Inter Ocean*, 26 July 1877, 1; Anthony Giddens, *Modernity and Self-Identity: Self and Society in the Late Modern Age* (Palo Alto, Calif.: Stanford University Press, 1991), 25–27.

43. "The Anxiety of the People," *New York Times*, 17 Aug. 1881, 1. Western Union provided free telegraphic bulletins of Garfield's medical condition, as it did a few years later when Ulysses Grant died after a long illness. Norvin Green to Postmaster General John Wanamaker, 16 July 1889, PLB WUTC.

44. "At the Fifth-Avenue Hotel. Eagerly Awaiting the News—Discussions in the Anxious Throng," *New York Times*, 3 July 1881, 2; "Brooklyn Much Disturbed," *New York Times*, 3 July 1881, 2; "The President's Condition," *New York Times*, 3 July 1881, 7; "The Scenes in Wall-Street," *New York Times*, 3 July 1881, 2; "General Telegraph News," *New York Times*, 7 July 1881, 2.

45. Col. E. A. Calkins, "Editing a Newspaper as an Art," in *Proceedings of the Wisconsin Editorial Association, Held at La Crosse, June 17, 1873* (Madison, Wis.: Atwood & Culver, 1874).

46. Royster, *Destructive War*, 233 and 241; Richard Menke, "Media in America, 1881: Garfield, Guiteau, Bell, Whitman," *Critical Inquiry* 31 (2005): 638–64.

47. See, for example, Richard D. Brown, *Knowledge Is Power: The Diffusion of Information in Early America, 1700–1865* (New York: Oxford University Press, 1989), 279–80.

48. Charles Richard Weld, *A Vacation Tour in the United States and Canada* (London: Longman, Brown, Green, and Longmans, 1855), 379; William Henry Charlton,

Four Months in North America (Hexham: J. Catherall, 1873), 28–29. For examples of hotels being telegraphic centers, see telegram from Thomas T. Eckert to William H. Seward, 12 Apr. 1864; and Edward S. Sanford to Seward, undated 1865, both in Seward Papers; "The News in This City," *New York Times*, 3 July 1881, 1.

49. Giddens, *Modernity and Self-Identity*, 25–27; James Russell Lowell, "Scotch the Snake, or Kill It?" *North American Review*, July 1865, 190; Rollo Ogden, "The Press and Foreign News," *Atlantic Monthly* 86 (Sept. 1900): 390–93; Shaw, "At the Crossroads," 41.

50. Sam Dillon, "In Test, Few Students Are Proficient Writers," *New York Times*, 3 Apr. 2008, xx; "Writing, Technology and Teens," Pew Internet and American Life Project report, 24 Apr. 2008, www.pewinternet.org/PPF/r/247/report_display.asp, accessed 8 Feb. 2009; Geoffrey Nunberg, "All Thumbs," 10 July 2008, http://people .ischool.berkeley.edu/~nunberg/texting.html, accessed 8 Feb. 2009.

51. See, for example, Laura Otis, *Networking: Communicating with Bodies and Machines in the Nineteenth Century* (Ann Arbor: University of Michigan Press, 2001); Paul Gilmore, "Aesthetic Power: Electric Words and the Example of Frederick Douglass," *American Transcendental Quarterly* 16 (2002): 291–92; Richard Menke, "Telegraphic Realism: Henry James's *In the Cage*," *PMLA* 115 (2000): 976–77.

52. Edmund Wilson, *Patriotic Gore: Studies in the Literature of the American Civil War* (New York: Oxford University Press, 1962), 635–69; Russell B. Nye, *Society and Culture in America, 1830–1860* (New York: Harper & Row, 1974), 94–95, 143–44; Thomas Gustafson, *Representative Words: Politics, Literature, and the American Language, 1776–1865* (Cambridge: Cambridge University Press, 1992); David Simpson, *The Politics of American English, 1776–1850* (New York: Oxford University Press, 1986); Kenneth Cmiel, *Democratic Eloquence: The Fight over Popular Speech in Nineteenth-Century America* (Berkeley: University of California Press, 1991).

53. Schwartzlose, *Nation's Newsbrokers*, 1:137; Richard B. Kielbowicz, "News Gathering by Mail in the Age of the Telegraph: Adapting to a New Technology," *Technology and Culture* 28 (1987): 37.

54. Jerusha Hall McCormack, "Domesticating Delphi: Emily Dickinson and the Electro-Magnetic Telegraph," *American Quarterly* 55 (2003): 569–601; Steve Paul, "Heretofore Unpublished Hemingway Observations Give Character, Sense of Place to KC," *Kansas City Star*, 27 June 1999, www.kcstar.com/hemingway/ehkc.shtml, accessed 22 May 2011; Shirer, *20th Century Journey*, 1:283–84.

55. Kenneth Cmiel, "'Broad Fluid Language of Democracy': Discovering the American Idiom," *Journal of American History* 79 (1992): 913.

56. Cmiel, *Democratic Eloquence*, 63, 65, 69, and 95; Anne C. Rose, *Voices of the Marketplace: American Thought and Culture, 1830–1860* (Oxford: Rowman & Littlefield, 1995), xx; Glen M. Johnson, ed., *The Topical Notebooks of Ralph Waldo Emerson* (Columbia: University of Missouri Press, 1994), 3:155.

57. Rufus Wilmot Griswold, ed., *The Prose Writers of America* (Philadelphia: Porter and Coates, 1854), 14; "Introduction: The Democratic Principle—The Importance of Its Assertion, and Application to Our Political System and Literature," *United States Magazine and Democratic Review* 1 (Oct. 1837): 14–15.

58. "The Union of Languages," *Journal of the Telegraph* 2, 15 Apr. 1869, 114; untitled, *Journal of the Telegraph* 2, 1 May 1869, 127.

Chapter 4. *"The Ticker Is Always a Treacherous Servant":* The Telegraph and the Rise of Modern Finance Capitalism

1. Edmund Clarence Stedman, ed., *The New York Stock Exchange* (New York: The Stock Exchange Historical Co., 1905), v–vi.

2. On the Rothschilds, see Niall Ferguson, *The House of Rothschild: Money's Prophets, 1798–1848* (New York: Penguin Books, 1998), 5. On Briggs's telegraph, see Geoffrey Wilson, *The Old Telegraphs* (London: Phillimore, 1976), 215.

3. As economic historian Alexander Field claims, the use of tickers and private-wire networks was "a quantum leap forward in the capacity of the American securities exchange system" and created "a trading regime that lasted roughly a century, from the 1870s through the 1960s." Field, "The Telegraphic Transmission of Financial Asset Prices and Orders to Trade: Implications for Economic Growth, Trading Volume and Securities Market Regulation," *Research in Economic History* 18 (1998): 177.

4. Robert Sobel, *Inside Wall Street* (New York: W. W. Norton, 1977), 27–28. For a description, see James K. Medberry, *Men and Mysteries of Wall Street* (1870; repr., New York: Greenwood Press, 1968), 20–38; Field, "Telegraphic Transmission of Financial Asset Prices," 164–65; Richard Sylla and George David Smith, "Information and Capital Market Regulation in Anglo-American Finance," in *Anglo-American Financial Systems: Institutions and Markets in the Twentieth Century*, ed. Michael D. Bordo and Richard Sylla (New York: Irwin Professional Publishing, 1995), 203; Caitlin Zaloom, *Out of the Pits: Traders and Technology from Chicago to London* (Chicago: University of Chicago Press, 2006), 165.

5. John J. McCusker, "The Demise of Distance: The Business Press and the Origins of the Information Revolution in the Early Modern Atlantic World," *American Historical Review* 110 (2005): 295–321; Bradford Scharlott and Mary Carmen Cupito, "Proto-Broadcasting in Cincinnati, 1847–1875: The Flow of Telegraph News to Merchants and the Press," *Ohio Valley History* 7 (2007): 32–46; Alexander Jones, *Historical Sketch of the Electric Telegraph, including Its Rise and Progress in the United States* (New York: George P. Putnam, 1852), 122–29.

6. See, for example, Kenneth D. Garbade and William L. Silber, "Technology, Communication and the Performance of Financial Markets, 1840–1975," *Journal of Finance* 33 (1978): 819–32; and Richard B. DuBoff, "Business Demand and the Development of the Telegraph in the United States, 1844–1860," *Business History Review* 54 (Winter 1980): 459–79.

7. The telegraph's most important economic effects were to reduce risk, informational asymmetry, and transaction costs. Alfred D. Chandler, *The Visible Hand: The Managerial Revolution in American Business* (Cambridge: Harvard University Press, 1977); James R. Beniger, *The Control Revolution: Technological and Economic Origins of the Information Society* (Cambridge: Harvard University Press, 1986); DuBoff, "Business Demand and the Development of the Telegraph in the United States";

William Cronon, *Nature's Metropolis: Chicago and the Great West* (New York: W. W. Norton, 1991); JoAnne Yates, *Control through Communication: The Rise of System in American Management* (Baltimore: Johns Hopkins University Press, 1989).

"The Telegraph," *DeBow's Review* 16 (1854): 168. Green's testimony is in House Committee on Post Offices and Post Roads, *Postal Telegraph Facilities*, 51st Cong., 1st sess., 1890, 41. Green also noted that New York City pool rooms, or off-track betting parlors, annually paid Western Union $750,000 to transmit horse racing odds and results.

8. Richard Franklin Bensel, *Yankee Leviathan: The Origins of Central State Authority in America, 1859–1877* (Cambridge: Cambridge University Press, 1990), chap. 4.

9. Medberry, *Men and Mysteries of Wall Street*, 10–11; James D. Reid, *Telegraph in America* (New York: Weed, Parsons, 1879), 484–85; George Kennan to family, 7 Oct. 1863, George Kennan Papers, Manuscript Division, Library of Congress, Washington, D.C.; entries of 17 Apr. and 29 May 1864, David Homer Bates Diary, Manuscript Division, Library of Congress, Washington, D.C.; Donald E. Markle, *Telegraph Goes to War: The Personal Diary of David Homer Bates, Lincoln's Telegraph Operator* (Lynchburg, Va.: Schroeder Publications, 2005), 112, n. 80. On Mullarkey, see William R. Plum, *Military Telegraph during the Civil War in the United States* (Chicago: Jansen, McClurg, 1882), 1:298, 307, 311. Recall that Col. William Gross of the USMT received a gift of stock in the South Western Telegraph Company from its president, Norvin Green; Green to Gross, 31 July and 27 Oct. 1865, William L. Gross Papers, Western Reserve Historical Society, Cleveland, Ohio. On southern operators engaging in speculation, see J. R. Dowell to W. S. Monteith, 19 Oct. 1863; David O'Keefe to Dowell, 22 Oct. 1863; Monteith to Dowell, 14 Nov. 1863; J. B. Tree to William S. Morris, 31 Oct. 1864, all in Southern Telegraph Co. Papers.

10. This summary of early ticker technology and business operations is drawn from *TAE Papers*, 1:602–10, 617–21; Horace L. Hotchkiss, "The Stock Ticker," in Stedman, *New York Stock Exchange*, 433–41; and Paul Israel, *From Machine Shop to Industrial Laboratory: Telegraphy and the Changing Context of American Invention, 1830–1920* (Baltimore: Johns Hopkins University Press, 1992), 103–7, 121–28.

11. This description is based on Reid, *Telegraph in America*, 614–15.

12. Reid, *Telegraph in America*, 609–10; Daniel H. Craig's testimony in *Report of the Committee of the Senate upon the Relations between Labor and Capital* (Washington, D.C.: GPO, 1885), 2:1269–70; Norvin Green to Senator Nathaniel P. Hill, 7 Mar. 1884, PLB WUTC.

13. Orton to John Van Horne, 26 Nov. 1869, PLB WUTC.

14. Annual Report to Directors, 22 Sept. 1873, G&S BoD; Orton to Stager, 12 Nov. 1870 and 24 Jan. 1871, PLB WUTC.

15. Meetings of 20 May and 4 Sept. 1871, G&S BoD; Orton to Anson Stager, 26 May 1869; Orton to John Van Horne, 26 Nov. 1869; Orton to Stager, 12 Nov. 1870; Orton to Marshall Lefferts, 8 Dec. 1870; Orton to James Simonton, 12 June 1871; Orton to Lefferts, 5 Sept. 1872, all PLB, WUTC.

16. M. D. Fagen, ed., *A History of Engineering and Science in the Bell System: The Early Years, 1875–1925* (New York: Bell Telephone Laboratories, 1975), 727–32.

17. G. L. Wiley to Theodore N. Vail, 17 Oct. 1881, AT&T Archives, Warren, N.J.

18. Reid, *Telegraph in America*, 621–26, 633–36; *TAE Papers*, 1:77–83, 163–72, 212–20; Edward A. Calahan, "The District Telegraph," *Electrical World and Engineer*, 16 Mar. 1901; "Memo. as to proposition of the Bell Telephone Co.," 10 June 1879, AT&T Archives; meetings of 20 July and 22 Dec. 1881, G&S ExCom; Henry C. Fischer and William H. Preece, "Joint Report upon the American Telegraph System," 511–15, British Post Office Archives, London.

19. Reid, *Telegraph in America*, 616–17; Fischer and Preece, "Joint Report upon the American Telegraph System," 207–19; *Relations between Labor and Capital*, 1:870–71; Reid, *Telegraph in America*, 2d ed. (New York: John Polhemus, 1886), 731; Statistical Notebooks, WUTC; "Productivity and Displacement of Labor in Ticker Telegraph Work," *Monthly Labor Review* 34 (1932): 1269–77.

20. House Judiciary Committee, *Telegraph Lines*, 43d Cong., 2d sess., 1875, H. Rep. 125; House Select Committee on Postal Telegraph, *Postal Telegraph in the United States*, 41st Cong., 2d sess., 1870, H. Rep. 114, 101–2; House Judiciary Committee, *In the Matter of the Western Union Telegraph Co.*, unpublished hearings, 43d Cong., 1st sess., 1875, (43) HJ.T2. On G&S's refusal to make exclusive local contracts, see meeting of 15 Aug. 1877, G&S ExCom.

21. Meetings of 7, 20, 29 May 1873, and 15 Aug. 1877, all in G&S ExCom; J. Wyman Jones to Chapman, 28 May 1873, Manhattan Quotation Telegraph Co. Records, NYSE; Hotchkiss, "Stock Ticker," 439–40.

22. Statistical Notebooks, WUTC.

23. Green to William F. Vilas, 17 Nov. 1887, PLB WUTC.

24. Elliot Gorn, *The Manly Art: Bare-Knuckle Prize Fighting in America* (Ithaca: Cornell University Press, 1986), 94; "Pastimes," *Chicago Daily Tribune*, 30 July 1876, 3; C. S. H. Small to Frank B. Rae, 21 June 1888, Frank B. Rae Incoming Correspondence, WUTC.

25. "Possible Advantages of a Telephone and Telegraph (W.U.) Combination," undated 1909 memo, AT&T Archives.

26. John Landesco, "Organized Crime in Chicago," in *The Illinois Crime Survey* (Chicago: Illinois Association for Criminal Justice, 1929), chap. 19. For an exposé on pool room gambling and Western Union's involvement with it, see Josiah Flynt's five-part series in *Cosmopolitan Magazine* in its February to June 1907 issues. Senate Special Subcommittee to Investigate Organized Crime in Interstate Commerce, *Investigation of Organized Crime in Interstate Commerce*, 81st Cong., 2d sess., 1950, 5–19.

27. Alexander Field makes this point in "Telegraphic Transmission of Financial Asset Prices."

28. Norvin Green to William H. Forbes, 19 Dec. 1879, PLB WUTC; WUTC Annual Reports, 1885, 5; 1887, 5; 1900, 8; 1901, 7; 1902, 7; 1903, 7; *Relations between Labor and Capital*, 1:876; Henry Clews, *Twenty-Eight Years in Wall Street* (New York: Irving, 1888), back matter; "Chicago Broker Expelled," *New York Times*, 14 Dec. 1899, 2; "Private Wire Contracts," Case 5421, *Interstate Commerce Commission Reports*, vol. 50 (Washington, D.C.: GPO, 1918), 738.

29. Kenneth Garbade and William L. Silber, "Technology, Communication, and the Performance of Financial Markets," *Journal of Finance* 33 (1978): 819–32; Henry Crosby Emery, *Speculation on the Stock and Produce Exchanges of the United States* (New York: Columbia University Press, 1896), 142; William Clarkson Van Antwerp, *The Stock Exchange from Within* (Garden City, N.Y.: Doubleday, Page, 1913), 284; Field, "Telegraphic Transmission of Financial Asset Prices," 164; Bradford Scharlott, "The Telegraph and the Integration of the U.S. Economy: The Impact of Electrical Communications on Interregional Prices and the Commercial Life of Cincinnati" (Ph.D. diss., University of Wisconsin, 1986), 121–23.

30. Sereno S. Pratt, *The Work of Wall Street* (New York: D. Appleton, 1910; reprint of 1903 edition), 133.

31. For general discussions of the effect of trading technologies upon traders' psychology, see Zaloom, *Out of the Pits*; Alex Preda, *Framing Finance: The Boundaries of Markets and Modern Capitalism* (Chicago: University of Chicago Press, 2009).

32. Hotchkiss, "Stock Ticker," 434; George Rutledge Gibson, *The Stock Exchanges of London, Paris, and New York: A Comparison* (New York: G. P. Putnam's Sons, 1889), 82–84; "Ticker Talk," *The Ticker* 2 (1908): 2 9; Susie M. Best, "A Song of the Ticker," *The Ticker*, Oct. 1910, 247.

33. Edward Neufville Tailer Diaries, New-York Historical Society, New York.

34. Emery, *Speculation on the Stock and Produce Exchanges*, 139; Statistical Notebooks, WUTC.

35. Richard Smitten, *Jesse Livermore: World's Greatest Stock Trader* (New York: John Wiley and Sons, 2001), 56–59; Edwin Lefevre, *Reminiscences of a Stock Operator* (New York: George H. Doran, 1923), 38.

36. Robert Sobel, *The Big Board: A History of the New York Stock Market* (New York: Free Press, 1965), 86–87; Clifford Browder, *The Money Game in Old New York: Daniel Drew and His Times* (Lexington: University Press of Kentucky, 1986), 100–101. Drew's grandson and a husband of one of his granddaughters were members of Groesbeck's firm.

37. Henry Demarest Lloyd, "Making Bread Dear," *North American Review* 137 (1883): 123–24. For a good description of how corners worked and their effects on the market, see William Cronon, *Nature's Metropolis: Chicago and the Great West* (New York: W. W. Norton, 1991), 127–32. Frank Norris's fictional account of the famous Leiter corner of 1897–98 conveys much of the glamour as well as avarice of grain speculation; Frank Norris, *The Pit: A Story of Chicago* (New York, Doubleday, Page, 1903).

38. Charles H. Taylor, *History of the Board of Trade of the City of Chicago* (Chicago: Robert O. Law, 1917), 739–59.

39. Lewis C. Van Riper, *The Ins and Outs of Wall Street* (n.p., 1898), 17; William Peter Hamilton, *The Stock Market Barometer* (New York: Harper & Brothers, 1922), 40, 144. For discussions of technical trading, see Zaloom, *Out of the Pits*, 170; and Alex Preda, "On Ticks and Tapes: Financial Knowledge, Communicative Practices, and Information Technologies in 19th Century Financial Markets," paper presented at Columbia Workshop on Social Studies of Finance, May 2002, copy in author's possession.

40. Medberry, *Men and Mysteries of Wall Street*, 122–23; Taylor, *History of the Board of Trade*, 565, 585, 1218–22; "Stock Exchange Reforms," *New York Times*, 14 Aug. 1887, 9.

41. Taylor, *History of the Board of Trade*, 586–88; "Glimpses of Gotham," *National Police Gazette*, 14 June 1879, 14; "The Spirit of Speculation. How the Passion for Gambling Is Being Fostered by Wall Street," *National Police Gazette*, 12 Feb. 1881, 11; "The Bucket-Shop Nuisance," *New York Times*, 4 Sept. 1884, 8. On defalcating clerks, see "A New-York Firm Robbed," 29 Oct. 1880, 5; "Another Cashier Astray," 19 June 1883, 1; "Ten Years in the Penitentiary," 18 Nov. 1883, 1, all in *New York Times*. On Livermore, see Lefevre, *Reminiscences of a Stock Operator*, 12–17; Cedric B. Cowing, *Populists, Plungers, and Progressives: A Social History of Stock and Commodity Speculation, 1890–1936* (Princeton: Princeton University Press, 1965), 101–3; Smitten, *Jesse Livermore*, 22–30.

42. "Women as Speculators," *New York Times*, 11 Sept. 1882, 8; "The Bucket-Shop Nuisance," *New York Times*, 4 Sept. 1884, 8.

43. On the so-called Big Four, see "Bucket Shop Sharpness," 3 Nov. 1887, 3; "Shops or Exchanges," 12 May 1889, 9, both in *New York Times*.

For details of bucket shop operations, including Haight and Freese, Coe, and M. J. Sage, see John Hill Jr., *Gold Bricks of Speculation: A Study of Speculation and Its Counterfeits, and an Expose of the Methods of Bucketshop and "Get-Rich-Quick" Swindles* (New York: Arno Press, 1975; reprint of Chicago: Lincoln Book Concern, 1904).; Merrill A. Teague, "Bucket-Shop Sharks," *Everybody's Magazine*, 14 and 15 (June–Sept. 1906): 723–35, 33–43, 245–54, and 398–408 ; *McCarthy v. Meaney*, 183 N.Y. 190 (1905); and "Bucket Shop Systems," *New York Times*, 19 Nov. 1905, 15; "Bucket Shops: An Inside View, and Some Lessons to Be Drawn Therefrom," *The Ticker* 2 (1908): 19–23; *Haight v. Haight & Freese*, 92 N.Y.S. 934 (1905).

44. Teague, "Bucket-Shop Sharks"; *Weiss v. Haight & Freese & Co.*, 148 Fed. 399 (1906). For similar cases involving other bucket shops, see *Joslyn v. Downing, Hopkins & Co.*, 150 Fed. 317 (1906); *Williamson v. Majors*, 169 Fed. 754 (1909).

45. Lefevre, *Reminiscences of a Stock Operator*, 15–16; *New York World*, 16 Dec. 1886, Mitchell Scrapbooks, NYSE.

46. Smitten, *Jesse Livermore*, 31–38.

47. Lefevre, *Reminiscences of a Stock Operator*, 19–20.

48. For a good account, see Jonathan Ira Levy, "Contemplating Delivery: Futures Trading and the Problem of Commodity Exchange in the United States, 1875–1905," *American Historical Review* 111 (2006): 307–35.

49. "The Bucket-Shop Nuisance," 4 Sept. 1884, 8; "Stock Exchange Quotations," 6 Mar. 1885, 8; "Bad for the Bucket Shops," 16 Apr. 1885, 1; "War on Bucket Shops," 5 Dec. 1886, 1; "All Tickers Ordered Out," 1 June 1889, 1; "Chicago's Business," 7 Jan. 1901, 15; "Haight & Freese Co. in Receivers' Hands," 10 May 1905, 1; "The Index Man of the Underworld of Finance," 18 Oct. 1908 (Sunday Magazine), 6; and "New Bucket Shops," 18 May 1913, 9, all in *New York Times*; Taylor, *History of the Board of Trade*, 787; *New York World*, 16 Dec. 1886; *New York Evening Post*, 1 June 1889, both in Mitchell Scrapbooks, NYSE; W. G. Nicholas, *Cold Facts about Bucket Shops* (Chicago: Business Publishing

Co., 1887), 2–6. New York Stock Exchange trading volumes can be found on its Web site, www.nyse.com, accessed 17 Mar. 2006.

50. "Tapping the Wires," 18 Oct. 1883, 1; "The 'Big Four' Suspends," 20 May 1890, 8; both in *New York Times*; Gold and Stock Telegraph Company attorney to George W. Ely, 25 March 1884, and James H. Milne to Ely, 4 Aug. 1884, both in Committee on Arrangements, General Files, NYSE; William P. Grovesteen to Governing Committee, 28 Feb. 1887, reproduced in "Stock Quotations. The New York Stock Exchange Should Control Its Own Quotations," undated pamphlet in Mitchell Scrapbooks, NYSE; memorandum giving list of bucket shops in New York City, undated 1886 or 1887, Officers' Files, Vice President R. H. Thomas, NYSE; "Tottering Bucket Shops," *New York Tribune*, 27 Apr. 1890, Mitchell Scrapbooks, NYSE.

51. Record Book of Louisville Office, 1877–1940, WUTC; "Bucket Shops Cut Off By Western Union," *New York Times*, 8 July 1910, 1; "Digest of the Preliminary Work of the Special Committee of June 25, 1913," Special Committee on Bucket Shop Operations, NYSE.

52. "Bucket Shops and Board of Trade," *Western Rural* 21 (17 Feb. 1883): 52; Hutchinson's quote is from Jonathan Lurie, *The Chicago Board of Trade, 1859–1905: The Dynamics of Self-Regulation* (Urbana: University of Illinois Press, 1979), 91; "War on the Bucket Shops," *New York Times*, 30 Aug. 1887, 2; *Springs v. James*, 121 N.Y.S. 1054 (1910).

53. Emery, *Speculation on the Stock and Produce Exchanges*, 98; *Report of Governor's Committee on Speculation in Securities and Commodities* (Albany, N.Y., 1909), 4 and 7.

54. *Haight & Freese's Guide to Investors* (Philadelphia: Haight & Freese, 1899), 25 and 58; C. C. Christie, "Bucket Shop vs. Board of Trade," *Everybody's Magazine* 15 (1906): 708–9 and 713.

55. "The Chicago Board of Trade, How It Helps the Farmer, Grain Dealer and Shipper," *The Ticker* 2 (1908): 255–60; Jackson Lears, *Something for Nothing: Luck in America* (New York: Penguin, 2004), 194; Ann Fabian, *Card Sharps, Dream Books, and Bucket Shops: Gambling in Nineteenth-Century America* (Ithaca, N.Y.: Cornell University Press, 1990), 4, 5, 157, 188, 198; Lurie, *Chicago Board of Trade*, 78–79.

56. *Public Grain and Stock Exchange v. Western Union Telegraph Co.*, 1 Ill. Cir. Ct. 548 (1883). Tuley's decision is quoted in *Bryant and another v. Western Union Tel. Co.*, 17 Fed. 825 (1883); *New York and Chicago Grain and Stock Exchange v. Chicago Board of Trade*, 127 Il. 157 (1889). For the board's reaction to the case, see Lurie, *Chicago Board of Trade*, 99–102, 138–51.

57. *Christie Grain and Stock Co. et al. vs. Board of Trade of City of Chicago*, 125 Fed. 161 (1903); *Board of Trade of City of Chicago v. L. A. Kinsey Co. et al.*, 125 Fed. 72 (1903); *Board of Trade of City of Chicago v. Donovan Commission Co. et al; Same v. Cella Commission Co. et al.*, 121 Fed. 1012 (1903); Taylor, *History of the Board of Trade*, 1218–22.

58. *Board of Trade of the City of Chicago v. Christie Grain and Stock Company and L.A. Kinsey Company v. Board of Trade of the City of Chicago*, 198 U.S. 236 (1905). Justices Harlan, Day, and Brewer dissented.

59. Norvin Green to Robert C. Clowry and Clowry to J. M. Ball, both 29 Dec. 1883, in meeting of 2 Jan. 1884, BoD CBOT.

60. "War on Bucket Shops," *New York Times*, 5 Dec. 1886, 1; Taylor, *History of the Board of Trade*, 787; Official Notification from Sec. George Stone to Western Union, 1 Mar. 1890; and contract between the Chicago Board of Trade, Western Union Telegraph Company, and the Postal Telegraph-Cable Company, 31 May 1892, both in BoD CBOT.

61. "Attacking Bucket Shops," *New York Times*, 20 Mar. 1897, 12; "Ending the Ticker," *New York Times*, 19 June 1897, 12; Committee on Arrangements, Minutes, 15 Jan. 1897, NYSE.

62. Meetings of 6 Nov. 1883; 29 Apr. 1890; 11 Sept. 1900; and 23 Mar. 1901, all in BoD CBOT. A sample application for ticker service is in meeting of 27 Jan. 1903, BoD CBOT.

63. Material on the Committee on Promotions and Council of Grain Exchanges is in Misc. Docs., CBOT. The Board of Trade's education and reform movement was akin to the efforts of the financial community to educate the public for the necessity of banking reform, which began in the late 1890s and culminated in the Federal Reserve Act of 1913; James Livingston, *Origins of the Federal Reserve System: Money, Class, and Corporate Capitalism, 1890–1913* (Ithaca, N.Y.: Cornell University Press, 1986), 33–34 and 71.

64. Lurie, *Chicago Board of Trade*, 161–63; Taylor, *History of the Board of Trade*, 903, 916, 976; Hill, *Gold Bricks of Speculation*, 355; "Stock Quotations Held Up," *New York Times*, 5 May 1909, 1; "Bucket Shops Open Here and Outside," *New York Times*, 16 May 1913, 20. On the Board of Trade's investigative activities, see material in Market Report Committee, Misc. Docs., CBOT.

65. "Federal Raids on Bucket Shops," 3 April 1910, 2; "Paul Lambert & Co. Fail," 19 Oct. 1915, 14; "Bucket-Shop Evil Soon To Be," 7 May 1916, E4, all in *New York Times*; USAG Annual Report for 1910, 23–24; USAG Annual Report for 1913, 46.

66. Law Committee Reports and Resolutions, 13 Feb. 1883, NYSE; "Counsel for a Bucket Shop," *New York Times*, 26 July 1893, 2.

67. John Phillip Quinn, *Fools of Fortune, or Gambling and Gamblers* (Chicago: Quinn Publishing Co., 1890), 577–78; *Board of Trade of City of Chicago v. O'Dell Commission Co. et al.*, 115 Fed. 574 (1902).

68. *Report of Governor's Committee on Speculation*, 3–5; Carl Parker, "Governmental Regulation of Speculation," *Annals of the American Academy of Political and Social Science* 38 (1911), 152; Van Antwerp, *Stock Exchange from Within*, 50; "Proceedings of the Sixth Annual Meeting of the Council of Grain Exchanges at Chicago, Illinois," 21 and 22 Jan. 1915, 2, Misc. Docs., CBOT.

69. Sereno Stansbury Pratt, *The Work of Wall Street: An Account of the Functions, Methods, and History of the New York Money and Stock Markets*, 2d ed. (New York: D. Appleton, 1912), 71.

70. Cowing, *Populists, Plungers, and Progressives*, 95; Steve Fraser, *Every Man a Speculator: A History of Wall Street in American Life* (New York: HarperCollins, 2005), 389; Lizabeth Cohen, *Making a New Deal: Industrial Workers in Chicago, 1919–1939* (New York: Cambridge University Press, 1990), 164, 175, 183–84; "Trading in Odd Lots New Market Force," *New York Times*, 30 Apr. 1916, 21.

71. Fraser, *Every Man a Speculator*, 577; Jeff Sommer, "Market Wisdom Applies to

E.T.F.'s, Too," *New York Times*, Sunday Business section, 2 Jan. 2011; Graham Bowley, "The New Speed of Money," *New York Times*, Sunday Business sec., 2 Jan. 2011.

Chapter 5. *"Western Union, by Grace of FCC and A.T.&T.":*
The Telegraph, the Telephone, and the Logic of Industrial Succession

1. The Bell Telephone Company was formed in the summer of 1877. The company set up local operating companies by assigning patent rights to territories in exchange for half the local companies' stock. In early 1879 it reorganized as the National Bell Telephone Company. After the 1879 agreement with Western Union, the parent company again reorganized as the American Bell Telephone Company in 1880 and brought its total capital up to $5 million. In 1885 the Bell interests established AT&T as the subsidiary of American Bell devoted to establishing the long-distance telephone business. In 1900 AT&T became the parent company of the Bell interests through an exchange of American Bell and AT&T stock and a rechartering of the corporation from Massachusetts to New York. For a good summary, see Gerald W. Brock, *The Telecommunications Industry: The Dynamics of Market Structure* (Cambridge: Harvard University Press, 1981), chap. 4.

2. In 1877 Western Union and Atlantic and Pacific agreed to pool their income and expenses at about the ratio of 85 to 15 percent respectively. In 1877 A&P handled about 2.5 million messages and Western Union about 21 million. WUTC Annual Report for 1877, 11; Henry C. Fischer and William H. Preece, "Joint Report upon the American Telegraph System," 471, British Post Office Archives, London; and James D. Reid, *Telegraph in America* (New York: Weed, Parsons, 1879), 577–90.

3. W. Bernard Carlson, "Entrepreneurship in the Early Development of the Telephone: How Did William Orton and Gardiner Hubbard Conceptualize the New Technology?" *Business and Economic History* 23 (1994): 169–70. On Gould and the A&P, see Maury Klein, *The Life and Legend of Jay Gould* (Baltimore: Johns Hopkins University Press, 1986), 195–205; "The American Rapid Telegraph Company" (1879); "The American Rapid Telegraph Company" (1880), both in WUTC; Reid, *Telegraph in America*, 2d ed. (New York: John Polhemus, 1886), 772–81. Gray's harmonic telegraph was capable of handling six simultaneous messages on short lines, but it proved touchy in actual operation, and Postal used it as a quadruplex between New York and Chicago for a few years. William Maver Jr., *American Telegraphy and Encyclopedia of the Telegraph* (New York: Maver, 1903), 355b.

4. Paul Israel, *From Machine Shop to Industrial Laboratory: Telegraphy and the Changing Context of American Invention, 1830–1920* (Baltimore: Johns Hopkins University Press, 1992), 128–51.

5. Orton to Anson Stager, 3 Feb. 1875; Orton to Nicholson, 8 Jan. 1878, both in PLB WUTC; Reid, *Telegraph in America* (1879 ed.), 564; Stephen B. Adams and Orville R. Butler, *Manufacturing the Future: A History of Western Electric* (Cambridge: Cambridge University Press, 1999), 24.

6. *Duplex* telegraphy refers to methods for transmitting two simultaneous messages on a telegraph line, one in each direction. Although there are various specific designs, the general principle of duplexing is the use of an "artificial line" to repro-

duce the electrical characteristics of the telegraph line on which it is installed. The receiving sounder is placed between the real and artificial lines, rendering it insensitive to the transmitting key. Stearns's key insight was to add capacitance as well as resistance to the artificial line.

Diplex telegraphy refers to methods for transmitting two simultaneous messages on a telegraph line, both in the same direction. The *quadruplex* is the combination of the duplex and diplex.

Multiplex is the general term for any method of transmitting more than one telegraph message simultaneously on one line.

7. *TAE Papers*, 1:513–15, 531–33, 555–56; 2:340–44.

8. In 1886 Western Union used 140 quadruplex instruments to create some 125,000 miles of additional circuits. At an approximate cost of $50 per mile to install new wires on existing poles, the quadruplex saved the company some $6 million by that year alone. Reid, *Telegraph in America* (1886 ed.), 676–78.

9. *TAE Papers*, 2:340–44.

10. Because of the lengthy litigation over the rights to the quadruplex, Edison never received direct payment from Western Union for the rights to the invention. In the fall of 1875, Edison and Western Union agreed to drop any monetary claims regarding the quadruplex and placed Edison on retainer to develop harmonic telegraph systems. The latter agreement also required that Edison use Western Union's patent attorney Grosvenor Lowrey for any patents resulting from this work. For a summary of the complex tangle surrounding the quadruplex, see *TAE Papers*, 2:312–14, 367, 378–80, 405, 506–7, 581–83, 691–97, 794–815. I also thank the director of the Edison Papers project, Paul Israel, for discussions on this point.

Although Gould and Edison reached formal agreement on 4 Jan. 1875, Gould claimed to have had control of Edison's quadruplex since early December 1874; see Gould to William E. Chandler, 5 Dec. 1874, William E. Chandler Papers, Manuscript Division, Library of Congress, Washington, D.C.

11. Bell filed his patent application on 14 Feb. 1876, a few hours before Elisha Gray filed his patent caveat on telephony. The crucial claim of Bell's patent was his claim to the transmission of sounds by means of "electrical undulations." Bell succeeded in transmitting speech about a month later using a method somewhat different from that shown in his patent. In January 1877 he filed a new patent for an improved telephone. Bernard S. Finn, "Bell and Gray: Just a Coincidence?" *Technology and Culture* 50 (2009): 193–201; Brock, *Telecommunications Industry*, 89–91; David A. Hounshell, "Elisha Gray and the Telephone: On the Disadvantages of Being an Expert," *Technology and Culture* 16 (1975): 133–61.

12. Robert V. Bruce, *Alexander Graham Bell and the Conquest of Solitude* (Boston: Little, Brown, 1973), 126–27, 137–50, 160–62.

13. Bell to Alexander Melville and Eliza Symonds Bell, 5 and 22 Mar. 1875, Bell Family Papers, Manuscript Division, Library of Congress, Washington, D.C.

14. Bruce, *Alexander Graham Bell*, 173–74; Rosario Joseph Tosiello, *The Birth and Early Years of the Bell Telephone System, 1876–1880* (New York: Arno Press, 1979), 223; *TAE Papers*, 2:524–26.

15. Matthew Josephson, *Edison: A Biography* (New York: McGraw-Hill, 1959), 141–42; Bruce, *Alexander Graham Bell*, 229–30; George David Smith, "Forfeiting the Future," *Audacity* 4 (1996): 24–39.

16. Bruce, *Alexander Graham Bell*, 229–30. See in particular Hubbard to Gertrude Hubbard, 16 Oct. 1876 and 15 Sept. 1877, Hubbard Family Papers, Manuscript Division, Library of Congress, Washington, D.C.; Hubbard to Cornish, 16 Sept. 1877; Hubbard to T. B. A. David, 14 Oct. 1877; and Watson to Hubbard, 4 Dec. 1877, all in AT&T Archives, Warren, N.J.

17. *Report of the Committee of the Senate upon the Relations between Labor and Capital, and Testimony Taken by the Committee* (Washington, D.C.: GPO, 1885), 1:882; Chauncey Depew, *My Memories of Eighty Years* (New York: C. Scribner's Sons, 1924), 176; William Chauncy Langdon, "Gardiner G. Hubbard's Offer of an Interest in the Telephone to Chauncey M. DePew" (1928), AT&T Archives; Thomas A. Watson, *The Birth and Babyhood of the Telephone* (New York: American Telephone and Telegraph, 1937), 23.

18. George Prescott, *Bell's Electric Speaking Telephone: Its Invention, Construction, Application, Modification and History* (New York: D. Appleton, 1884), 444; Robert W. Garnet, *The Telephone Enterprise: The Evolution of the Bell System's Horizontal Structure* (Baltimore: Johns Hopkins University Press, 1985), 11.

19. During Orton's tenure as president between 1866 and 1878, the company spent about $1 million on developing and acquiring telegraph and telephone patents, and his successor Norvin Green estimated that their value to Western Union was twenty to fifty times this amount. See WUTC Annual Report for 1869, 15 and 37–40; Orton to James D. Reid, 21 Sept. 1869; Green to Gen. Wager Swayne, 19 Mar. 1881, both in PLB WUTC. For Edison's work on the telephone, see Bruce, *Alexander Graham Bell*, 173–74; and Tosiello, *The Birth and Early Years of the Bell Telephone System*, 223.

20. "Congress and the Telegraph," *Telegrapher* 10 (1874): 134.

21. The claim that Orton's dislike for Hubbard led him to reject his offer apparently originated with Western Union electrician George Prescott. In his 1884 book on the telephone, Prescott claimed that in the summer of 1875 Bell briefly conducted telephone experiments in Western Union's headquarters, but when Orton learned that Hubbard, "who was personally obnoxious to him," was Bell's principal financial backer, Orton stopped Bell's experiments. However, Hubbard's papers show that he and Orton were on friendly terms and that they frequently socialized together. Rather than Orton, it seems that Western Union's directors were the source of antipathy for Hubbard. On 4 December 1877 Thomas Watson wrote Hubbard that one Western Union official had told him that "Mr. Hubbard had caused the W.U. a great deal of trouble and they were determined to sit down on him this time." Prescott, *Bell's Electric Speaking Telephone*, 444; Hubbard to Gertrude Hubbard, 25 Nov. 1877, 16 Dec. 1870, 2 Oct. 1873; and Hubbard to Robert McCurdy, 13 Feb. 1874; all in Hubbard Papers; Watson to Hubbard, 4 Dec. 1877, AT&T Archives.

22. Gray to A. L. Hayes, 2 Nov. 1876, Elisha Gray Collection, National Museum of American History Archives Center, Smithsonian Institution, Washington, D.C.; Fischer and Preece, "Joint Report upon the American Telegraph System," 396–98;

Hubbard to Thomas E. Cornish, 16 Sept. 1877; Hubbard to Thomas Watson, 18 Sept. 1877, both in AT&T Archives; Bruce, *Alexander Graham Bell*, especially chaps. 17–20.

23. Orton to William Hunter, 20 July 1877; Orton to T. B. A. David, 15 Aug. 1877, both in PLB WUTC; Hubbard to Cornish, 16 Sept. 1877; Hubbard to T. B. A. David, 14 Oct. 1877, both in AT&T Archives; Hubbard to Bell, 23 Sept. and 18 Oct. 1877, both in Bell Papers; Prescott, *Bell's Electric Speaking Telephone*, 444–45; Tosiello, *The Birth and Early Years of the Bell Telephone System*, 250–53.

24. Edison to Orton, 17 Sept. 1877, TAED LB001288; Pope to William Henry Preece, 22 Mar. 1878, TAED Z005AV; Tosiello, *The Birth and Early Years of the Bell Telephone System*, 225–28; Hubbard to Bell, 1 Feb. 1878, Bell Papers; Orton to Charles A. Cheever, 21 Feb. 1878; Orton to Samuel S. White, 1 Mar. 1878, both in PLB WUTC.

25. Hubbard later admitted that he "doubt[ed] if any instrument of value can be made under [Bell's] patents of 1876" and that Bell's patents of 1877 contained "the valuable features of the invention." Hubbard to Cornelius Roosevelt, 21 Jan. 1878; Hubbard to Gertrude Hubbard, 13 and 21 Sept. 1878, all in Hubbard Papers. For a general account, see Garnet, *Telephone Enterprise*, 32–33. On agent commissions, see Hubbard to Cornish, 26 Feb. 1878, AT&T Archives. The U.S. Supreme Court decision is *The Telephone Cases*, 126 U.S. 1 (1888).

26. On Western Union's acquisition of Gold and Stock, see Orton to Anson Stager, 22 Mar. 1871, and Executive Order 124, 15 June 1871, both in PLB WUTC; Israel, *From Machine Shop to Industrial Laboratory*, 125–27.

27. Green to George Gifford, 8 July 1879; Green to Charles A. Cheever, 9 Feb. 1878; "memorandum for Dr. White," May 1879, all in PLB WUTC; memo draft of July 1879, AT&T Archives.

28. Green to Forbes, 3 Sept. 1879, PLB WUTC; *TAE Papers*, 5:419–21; Alan Stone, *Public Service Liberalism: Telecommunications and Transitions in Public Policy* (Princeton: Princeton University Press, 1991), 65–66.

29. "Affidavit of George Gifford," 19 Sept. 1882, AT&T Archives; Green to Forbes, 3 Sept. 1879, PLB WUTC.

30. Gould's drive at Western Union unfolded in two stages. In 1874 he acquired a small competing telegraph company, the Atlantic and Pacific, and used his access to railroad rights-of-way to expand it into a major competitor. Western Union bought it out in late 1877, giving Gould a large block of Western Union stock. In May 1879 Gould set up the Central Union Telegraph Company. In August, after a significant expansion of its facilities, Gould unveiled it as the American Union. In early 1881 Gould acquired working control of Western Union, replacing William H. Vanderbilt as the shareholder of record. See Richard R. John, *Network Nation: Inventing American Telecommunications* (Cambridge: Harvard University Press, 2010), 166–70; Julius Grodinsky, *Jay Gould: His Business Career, 1867–1892* (Philadelphia: University of Pennsylvania Press, 1957), 148–62, 269–85; and Klein, *The Life and Legend of Jay Gould*, 195–205, 277–82.

31. David Homer Bates to Vail, 23 May and 26 June 1879; Vail to D. H. Ogden, 26 May 1879; Hall to Vail, 23 June 1879; Vail to J. J. Storrow, 24 Jan. 1889, all in AT&T Archives; and Tosiello, *The Birth and Early Years of the Bell Telephone System*, 460–67; J. Leigh Walsh,

Connecticut Pioneers in Telephony: The Origin and Growth of the Telephone Industry in Connecticut (New Haven, Conn.: Telephone Pioneers of America, 1950), 77–91.

32. Green to Gifford, 8 July and 18 Aug. 1879; Green to Sir Hugh Allan, 19 Aug. 1879; Green to W. H. Forbes, 3 Sept. 1879, all in PLB WUTC. For a general account, see Bruce, *Alexander Graham Bell*, 270–71. For Western Union's equity stakes in telephone exchanges, see WUTC Annual Report for 1880, 8–9.

33. WUTC Annual Report for 1880, 8–9; Green to James Demarest, 22 Mar. 1883; Green to Col. E. W. Cole and E. S. Babcock, 20 Aug. 1884; Green to Forbes, 6 Apr. 1887, all in PLB WUTC; Executive Committee Minutes, 24 Dec. 1902, WUTC; Annual Reports of American Speaking Telephone Company, WUTC; Green to Forbes, 6 Apr. 1887, AT&T Archives; Stone, *Public Service Liberalism*, 116–19; and George David Smith, *The Anatomy of a Business Strategy: Bell, Western Electric, and the Origins of the American Telephone Industry* (Baltimore: Johns Hopkins University Press, 1985).

34. J. J. Storrow to W. H. Forbes, 27 Feb. and 8 Mar. 1880; Forbes to Green, 31 Oct. and 10 Dec. 1881, 3 Mar. 1882; Storrow to Forbes, 6 and 8 Mar. 1880; Forbes to Green, 17 Feb. and 25 July 1885, all in AT&T Archives; Arthur S. Pier, *Forbes: Telephone Pioneer* (New York: Dodd, Mead, 1953), 153–57; Garnet, *Telephone Enterprise*, 80.

35. G&S Annual Reports for 1880 and 1881, WUTC; Record Book, American Speaking Telephone Co., entries for 22 Jan. 1883 and 14 Apr. 1913, WUTC; Forbes to Storrow, 2 Mar. 1883; E. R. Hoar to Forbes, 6 Mar. 1883; Vail to Storrow, 24 Jan. 1889; Charles H. Swan to Hutchinson, 26 Jan. 1901; J. H. Benton to AT&T, 5 Apr. 1905; "Introductory Statement," n.d., box 1094, all in AT&T Archives; Federal Communications Commission, *Investigation of the Telephone Industry in the United States*, 76th Cong., 1st sess., H.R. Doc. 340, 1939, 124; "Fight for Millions," *Boston Herald*, 14 June 1899, from "Misc. Clippings, Telephone," WUTC.

36. Green to Messers Winslows, 29 Nov. 1879; Green to Forbes, 7 Feb. 1880, both in PLB WUTC; Vail to Western Union, 6 Aug. 1880, AT&T Archives.

37. Lillian Hoddeson, "The Emergence of Basic Research in the Bell Telephone System, 1875–1915," *Technology and Culture* 22 (1981): 512–44; and Milton Mueller, "The Switchboard Problem: Scale, Signaling, and Organization in Manual Telephone Switching, 1877–1897," *Technology and Culture* 30 (1989): 534–60.

38. B. P. Hamilton, "Early Bell System Telegraph Services," *Bell Laboratories Record* 23 (1945): 373–76; Forbes to Green, 7 Mar. 1887; Green to Forbes, 6 Apr. 1887; Green to Stockton, 25 Feb., 17 Aug., and 30 Oct. 1888; Green to Hudson, 15 Jan. 1891, all in AT&T Archives.

39. Forbes to Stockton, 5 Apr. 1888, AT&T Archives; Vail to Frederick Fish, 14 Apr. 1906, quoted in FCC, *Investigation of the Telephone Industry*, 207. The FCC concluded in the same report (p. 232) that Thomas Lockwood had also favored a dominant position for Bell in the telegraph industry as early as 1888. M. D. Fagen, ed., *History of Engineering and Science in the Bell System: The Early Years, 1875–1925* (New York: Bell Telephone Laboratories, 1975), 733.

40. The operation of the composite circuit rests upon the insight that telegraph signals are essentially direct current and telephone signals are alternating current and thus do not interfere with each other over properly configured circuits. For a

technical discussion, see Fagen, *History of Engineering and Science in the Bell System*, 734–38.

41. Bell's assessments of the Van Rysselberghe system are in box 1048, AT&T Archives. Lockwood's reminiscences are from Frank Pickernell, "The Introduction of the Composite System," *Proceedings of the Telephone Pioneers of America*, sixth meeting (1916): 69–73. See also F. W. Dunbar, *The Present and Probable Future Relations between the Telegraph and Telephone* (n.p., 1895), 8. In the 1890s AT&T also investigated several printing telegraph systems; see reports on these systems in boxes 1264 and 1307 and loc. 632-02-06, AT&T Archives. For Bell's work on long-distance telephony at this time, see Neil Wasserman, *From Invention to Innovation: Long Distance Telephone Transmission at the Turn of the Century* (Baltimore: Johns Hopkins University Press, 1985).

During the 1890s, Bell also actively investigated at least three printing telegraphs for its use but decided against using them, because the real bottleneck was the collection and delivery of telegrams, not the speed of transmission over wires. Lockwood to Hudson, 3 Aug. 1893; W. W. Swan to Hudson, 30 Oct. 1893; Lockwood to Hudson, 30 Jan. 1895; Lockwood to Hudson, 17 Apr. 1900, all in AT&T Archives.

42. Statistical Notebooks, WUTC; Fagen, *History of Engineering and Science in the Bell System*, 744.

43. Green to A. W. Campbell, 29 Aug. 1889, PLB WUTC; *Western Electrician* 14 (26 May 1894): 262; Edward J. Hall to Hudson, 12, 15, and 24 Dec. 1892, all in AT&T Archives.

44. James M. Herring and Gerald C. Gross, *Telecommunications: Economics and Regulation* (New York: McGraw-Hill, 1936), 4; Albert C. Crehore and George O. Squier, "Telegraphic, Telephonic, and Mail Service in the United States," *Western Electrician* 19 (8 May 1897): 263; John C. Tomlinson to Vail, 18 May 1903, AT&T Archives; "Topics of the Times," *New York Times*, 31 May 1902, 8.

45. John Brooks, *Telephone: The First Hundred Years* (New York: Harper & Row, 1976), 120–26; FCC, *Investigation of the Telephone Industry*, 87–90; Albert Bigelow Paine, *In One Man's Life: Being Chapters from the Personal and Business Career of Theodore N. Vail* (New York: Harper Brothers, 1921), 245–50; Noobar Danielian, *AT&T: The Story of Industrial Conquest* (New York: Vanguard Press, 1939), 56–57; Jean Strouse, *Morgan: American Financier* (New York: Harper Perennial, 2000), 563.

46. The Mackay Companies trusteeship controlled some one hundred state-based operating companies that collectively constituted the Postal Telegraph-Cable system. Although this trusteeship was cumbersome to manage from an accounting perspective, it allowed Mackay to evade Interstate Commerce Commission reporting requirements until after World War I. Tomlinson to Vail, 18 May and 2 June 1903, AT&T Archives. For information on Postal's structure see material in RG 28, entry 28, box 303, National Archives, Washington, D.C.

47. Vail to Fish, 14 Apr. 1906, quoted in Danielian, *AT&T*, 56–57.

48. Brock, *Telecommunications Industry*, 152–53; Brooks, *Telephone*, 133–35.

49. M. C. Rorty to Belvedere Brooks, 15 Sept. 1913; "Joint Arrangements between

the Western Union and the Bell Telephone System as of September 1913"; Rorty to Carlton, 28 Jan. 1910, all in AT&T Archives.

50. Acquisition of Western Union's telephone holdings was an important motivation for AT&T. See Stone, *Public Service Liberalism*, 180–84; and C[lement]. M[elville]. Keys, "The Rulers of the Wires," *World's Work* 19 (Mar. 1910): 12726–29. On the benefits to AT&T, see "Possible Advantages of a Telephone and Telegraph (W.U.) Combination," undated 1909 memo, AT&T Archives; and supplementary report to memorandum from McKay to Gallaher, 8 Mar. 1933, 10–11, WUTC.

51. George P. Oslin, *The Story of Telecommunications* (Macon: Mercer University Press, 1992), 259 and 262; "Interview of Gen. Robb by Mr. Langdon" (transcript of telephone interview), 4 Aug. 1933; "Report of Price, Waterhouse & Co. Chartered Accountants on the Property and Business of the Western Union Telegraph Co.," 4 Oct. 1910, both in AT&T Archives.

52. H. A. Bullock, "Reorganizing the Wires," *Boston Evening Transcript*, 15 Apr. 1911, 3.

53. See WUTC Annual Reports between 1910 and 1914, and Oslin, *Story of Telecommunications*, 266. Boxes 10 and 59 in the AT&T Archives contain a great deal of material on joint telephone-telegraph arrangements with Western Union regarding shared offices and wire plant.

54. Kingsbury to James T. Moran, 28 Dec. 1911 and 14 Feb. 1912; "Interview of Gen. Robb," all in AT&T Archives; *Annual Report*, Mackay Companies (New York, 1912). For material on relations between AT&T and Postal in this period, see boxes 9 and 10, AT&T Archives.

55. Garnet, *Telephone Enterprise*, 152–54, is a good summary.

56. "Vail's Statement Rebutting Congressman Lewis," 20 Jan. 1914; Wilson to Kingsbury, 19 Feb. 1912, both in AT&T Archives; "Finds Leased Wire Abuses Persistent," *New York Times*, 3 July 1914, 10. On the importance of railroad rights-of-way for AT&T's future telegraph business, see the 1935 testimony of Frank C. Page, reproduced in hearings before the Senate Committee on Interstate Commerce on S. Res. 95, "To Authorize a Complete Study of the Telegraph Industry," 76th Cong., 1st sess., 22 and 23 May 1939, 37.

57. Agreements and correspondence on printing telegraph patents are in box 61, AT&T Archives; see specifically Rorty to Gifford, 8 Mar. 1921 and "Memorandum for Mr. A. H. Griswold," 27 Oct. 1921. See also "Relationship of AT&T Co. to Telegraph Business," internal memo, Chester McKay to Mr. Gallaher, 8 Mar. 1933, 9–10, and supplementary report to memo from McKay to Gallaher, 8 Mar. 1933, both in WUTC.

58. For financial figures, see WUTC Annual Reports for 1914, 1929, and 1930; and Carrie Glasser, "Some Problems in the Development of the Communications Industry," *American Economic Review* 35 (1945): 592, 597, 599. For the 1929 and 1933 revenue figures, see testimony of Frank C. Page, 38. Quotes are from Oslin, *Story of Telecommunications*, 319–20; and statement of FCC Chairman James Lawrence Fly in "Study of the Telegraph Industry," Hearings before the Committee on Interstate Commerce, U.S. Senate, 77th Cong., 1st sess., 19–29 May 1941, 85.

59. Figures are from Fly's statement (pp. 10–11) and supplementary report to memo from McKay to Gallaher, 8 Mar. 1933, WUTC.

60. Stevenson to Thayer, 31 Aug. 1921, AT&T Archives. See also James M. Herring, "Public versus Private Ownership and Operation of the Communication Utilities," *Annals of the American Academy of Political and Social Science* 201 (Jan. 1939): 99; Oslin, *Story of Telecommunications*, 304–5, 320; and supplementary report to memo from McKay to Gallaher, 8 Mar. 1933, 14–15. For details on Bell's decision to acquire Teletype Corp., see C. G. Stoll to Edgar S. Bloom, 13 Aug. 1930; Bloom to Walter Gifford, 14 Aug. and 11 Sept. 1930; Frank Jewett to C. P. Cooper, 21 Aug. 1930, all in AT&T Archives.

61. Rorty to Walter Gifford, 8 Mar. 1921, AT&T Archives.

62. "Relationship of AT&T Co. to Telegraph Business," 5–6; M. C. Rorty to Walter Gifford, 18 Mar. 1921; J. J. Pilliod to F. A. Stevenson, 30 Aug. 1921, both in AT&T Archives. See also Hyman Howard Goldin, "The Domestic Telegraph Industry and the Public Interest: A Study in Public Utility Regulation" (Ph.D. diss., Harvard University, 1950), 226–36.

63. "Relationship of AT&T Co. to Telegraph Business," 5–6; Danielian, *AT&T*, 166.

64. D'Humy to White, 12 Aug. 1935, 16 Dec. 1937, and 15 April 1938, all in WUTC; Robert R. Mullen, "Send Your Own Telegram," *Christian Science Monitor*, Weekly Magazine Section, 3 Feb. 1940, 8–9; Oslin, *Story of Telecommunications*, 319–20.

65. "Relationship of AT&T Co. to Telegraph Business," 10–11; Herring and Gross, *Telecommunications*, 193; Danielian, *AT&T*, 165.

66. Fly's statement, 20.

67. President's Communications Policy Board, *Telecommunications: A Program for Progress* (Washington, D.C.: GPO, 1951), 95; "A Special Report to the Stockholders and Employees of the Western Union Telegraph Company on the Future of the Record Communications Industry," 20 Oct. 1949, AT&T Archives; "Comments of George L. Best Regarding the Western Union Telegraph Company's Statement to the President's Communications Policy Board on November 27, 1950," AT&T Archives.

68. "Western Union Hums—With Data," *Business Week*, 20 Feb. 1965, 151; T. A. Wise, "Western Union, by Grace of FCC and A.T.&T.," *Fortune*, March 1959, 114–19, 217–28.

69. "Private-Wire Network Connects 188 U.S. Banks," *Business Week*, 16 Sept. 1950, 106; "Western Union, by Grace of FCC and A.T.&T."

70. "Electronics Puts Young Blood in Old Company," *Business Week*, 27 Aug. 1960, 87–92.

71. "Western Union Hums—With Data"; "Uneven Match?" *Forbes*, 15 Mar. 1968, 81–82; "FCC Studies Western Union Bids for Satellite Network," *Aviation Week and Space Technology*, 16 Nov. 1966, 33.

72. "Western Union Finds a Connection," *Business Week*, 25 May 1968, 138–40.

73. "Lazarus," *Forbes*, 1 Oct. 1972, 26–27; "Western Union in a Tightening Squeeze," *Business Week*, 8 June 1974, 68–75.

74. "This Time, Maybe?" *Forbes*, 3 Mar. 1980, 82.

75. "Losing a Satellite Will Cost Western Union Plenty," *Business Week*, 20 Feb.

1984; Maggie Mahar, "Is Someone Sending a Message?" *Barron's*, 22 Dec. 1986, 13; "Drexel's Heavy Hand," *Newsweek*, 23 Nov. 1987, 53–55.

76. Robert Roy Britt, "Era Ends: Western Union Stops Sending Telegrams," 1 Feb. 2006, www.livescience.com/6989-era-ends-western-union-stops-sending-telegrams .html, accessed 7 Apr. 2012; Verlyn Klinkenborg, "The Telegram," *New York Times*, 8 Feb. 2006.

77. See, for example, Robert Sobel, *When Giants Stumble: Classic Business Blunders and How to Avoid Them* (Paramus, N.J.: Prentice Hall Press, 1999); Paul Solman and Thomas Friedman, *Life and Death on the Corporate Battlefield: How Companies Win, Lose, Survive* (New York: Simon and Schuster, 1982); Clayton M. Christensen, *The Innovator's Dilemma: When New Technologies Cause Great Firms to Fail* (Boston: Harvard Business School Press, 1997); Kevin Kennedy and Mary Moore, *Going the Distance: Why Some Companies Dominate and Others Fail* (New York: Prentice Hall, 2003); and Kenneth Labich and Patty de Losa, "Why Companies Fail," *Fortune* 130 (14 Nov. 1994): 22–32.

Some later scholarship calls for managers and organizational theorists to study rather than shun the subject of corporate failure. See a 2005 special issue of *Long Range Planning*; Tom McGovern, "Why Do Successful Companies Fail? A Case Study of the Decline of Dunlop," *Business History* 49 (2007): 886–907; and Howard Stanger, "Failing at Retailing (and Other Things): The Decline of the Larkin Company, 1918–1941," *Business and Economic History Online*, 2008, www.h-net.org/~business/bhc web/publications/BEHonline/2008/program08.html.

78. Andrew S. Grove, *Only the Paranoid Survive: How to Exploit the Crisis Points that Challenge Every Company* (New York: Broadway Business, 1999), 3–5.

79. T. J. Stiles, *The First Tycoon: The Epic Life of Cornelius Vanderbilt* (New York: Alfred A. Knopf, 2009), 510–11, 553–54; Klein, *Life and Legend of Jay Gould*, 195–205, 276–82.

80. For a classic analysis, see Theodore Levitt, "Marketing Myopia," *Harvard Business Review* 38 (1960): 45–56.

81. For a general analysis of competition between incumbent and upstart technologies, see Harmeet Sawhney and Xiaofei Wang, "Battle of Systems: Learning from Erstwhile Gas-Electricity and Telegraph-Telephone Battles," *Prometheus* 24 (2006): 235–56.

82. http://email.about.com/od/emailtrivia/f/emails_per_day.htm; www.phonedog .com/cell-phone-research/blog/offbeat-news-americans-send-4-1-billion-text-mes sages-daily.aspx, both accessed 5 Jan. 2010.

Conclusion. *The Promise of Telegraphy*

1. Dan Joling, "Global Warming Opens Arctic for Tokyo-London Undersea Cable," *Seattle Times*, 21 Jan. 2010, at http://seattletimes.nwsource.com/html/nation world/2010852730_apusarcticcable.html, accessed 15 Jan. 2011.

2. House Committee on Commerce, "Electromagnetic Telegraphs," U.S. Government Documents Serial Set, 25th Cong., 2d sess., H.R. Rep. 753, 6 Apr. 1838.

3. Henry David Thoreau, *Walden; or, Life in the Woods* (Boston: Houghton, Mifflin, 1897), 1:84–85.

4. As communications scholar John Durham Peters phrases it, "in principle" the telegraph reduced the "coefficient of friction for signals" to zero. See John Durham Peters, *Speaking into the Air: A History of the Idea of Communication* (Chicago: University of Chicago Press, 1999), 138–39, 184–85. For a more critical discussion of the potential versus actual uses of new communication technologies, see Paul Duguid, "Material Matters: Aspects of the Past and Futurology of the Book," in Geoffrey Nunberg, ed., *The Future of the Book* (Berkeley: University of California Press, 1996), 63–102.

5. Jeffrey Kieve, *Electric Telegraph: A Social and Economic History* (Newton Abbot: David & Charles, 1973); Russell W. Burns, *Communications: An International History of the Formative Years* (London: Institution of Electrical Engineers, 2004), 113–15.

6. William Clarkson Van Antwerp, *Stock Exchange from Within* (Garden City, N.Y.: Doubleday, Page, 1913), 148–50, 284, 341–48, 388; John C. Coffee Jr., "The Rise of Dispersed Ownership: The Role of Law in the Separation of Ownership and Control," Columbia Law School, Working Paper No. 182, Jan. 2001; Ranald C. Mitchie, *The Global Securities Market: A History* (Oxford: Oxford University Press, 2006).

7. Wolfgang Schivelbusch pioneered the study of the psychological effects of technological change in *The Railway Journey: Trains and Travel in the 19th Century*, trans. by Anselm Hollo (New York: Urizen Books, 1979). See also the following recent works: John Henry Hepp IV, *The Middle-Class City: Transforming Space and Time in Philadelphia, 1876–1926* (Philadelphia: University of Pennsylvania Press, 2003); Joy Parr, *Sensing Changes: Technologies, Environments, and the Everyday, 1953–2003* (Vancouver: UBC Press, 2010); Mats Fridlund, "The Power of Things: Towards a Phenomenological History of Technology," paper presented at Society for the History of Technology conference, Tacoma, Wash., Oct. 2010. See also a work in progress by Jonas Harvard, "Distant News and Local Opinion: How the Telegraph Affected Spatial and Temporal Horizons in Northern Scandinavia, 1850–1880," at http://nordicspaces.com/distant-news/, accessed 23 Dec. 2010.

8. Claude S. Fischer, *America Calling: A Social History of the Telephone to 1940* (Berkeley: University of California Press, 1992).

9. For a critical account of technological determinism, see David E. Nye, *Technology Matters: Questions to Live With* (Cambridge, Mass.: MIT Press, 2006), chaps. 2–4. Recently several other scholars have urged the field to renew interest in how technology drives social change. See, for example, Robert L. Heilbroner, "Technological Determinism Revisited," in Merritt Roe Smith and Leo Marx, eds., *Does Technology Drive History? The Dilemma of Technological Determinism* (Cambridge, Mass.: MIT Press, 1994), 67–78; Paul Ceruzzi, "Moore's Law and Technological Determinism: Reflections on the History of Technology," *Technology and Culture* 46 (July 2005): 584–93; Wiebe E. Bijker, "Globalization and Vulnerability: Challenges and Opportunities for SHOT around Its Fiftieth Anniversary," *Technology and Culture* 50 (July 2009): 610; Steven W. Usselman, "From Sputnik to SCOT: The Historiography of American Technology," *OAH Magazine of History* 24 (July 2010): 13.

The main purpose of this essay is to acquaint readers with the primary and secondary sources I have used to research and write this study. In that sense, it fulfills the purpose of the traditional scholarly bibliography. This essay, however, complements the endnotes instead of reproducing them—not all sources cited in the endnotes appear here, and sources not mentioned in the endnotes are discussed in this essay. I have chosen this format because I hope that this will assist future scholars who wish to re-search the history of telegraphy more fully. Thus, I do not limit the essay on sources to the particular topics covered in this book, though I make no claim to completeness.

General Histories of Telegraphy

The two best histories of telegraphy are Richard R. John, *Network Nation: Inventing American Telecommunications* (Cambridge: Harvard University Press, 2010); and Robert Luther Thompson, *Wiring a Continent: The History of the Telegraph Industry in the United States, 1832–1866* (Princeton: Princeton University Press, 1947). John's book will remain for some time the definitive account of American wire communication during the century between Morse's invention of the telegraph in the late 1830s and the establishment of the Federal Communications Commission in the mid-1930s. Thompson's book is the standard treatment of the telegraph industry before 1866. Although short on analysis, it is thoroughly researched and very detailed. It is not likely to be superseded. An excellent labor history of the telegraph industry that takes the 1883 operators' strike as its centerpiece is Edwin Gabler, *The American Telegrapher: A Social History, 1860–1900* (New Brunswick, N.J.: Rutgers University Press, 1988). Also quite valuable is Gregory J. Downey, *Telegraph Messenger Boys: Labor, Technology, and Geography, 1850–1950* (New York: Routledge, 2002). Although more relevant to ocean cables, a good history of telegraphy's effect on international relations is David Paull Nickles, *Under the Wire: How the Telegraph Changed Diplomacy* (Cambridge: Harvard University Press, 2003).

On the use of the telegraph by firms and businessmen, begin with Alfred D. Chandler's classic *The Visible Hand: The Managerial Revolution in American Business* (Cambridge: Harvard University Press, 1977). Other valuable work includes James R. Beniger, *The Control Revolution: Technological and Economic Origins of the Infor-*

mation Society (Cambridge: Harvard University Press, 1986); JoAnne Yates, *Control through Communication: The Rise of System in American Management* (Baltimore: Johns Hopkins University Press, 1989); and Gary Fields, *Territories of Profit: Communications, Capitalist Development, and the Innovative Enterprises of G. E. Swift and Dell Computer* (Stanford, Calif.: Stanford University Press, 2004).

The indispensable archival collection for the history of the American telegraph industry is the Western Union Telegraph Company records held at the Lemelson Center, Smithsonian Institution. This massive collection of more than eight hundred boxes is a necessity for research on virtually every facet of American telegraphy between its origins and its ultimate demise. Several parts of this collection have been important in researching this book. More than thirty letterbooks of presidents William Orton and Norvin Green contain copies of their official outgoing correspondence between 1866 and 1892. Their letters provide a wealth of information about the company's strategy and high-level decision making during this important period. One series of the collection contains the records of almost every telegraph company that Western Union bought out. These records trace the activities of competitors that entered the industry after 1866 and that Western Union later acquired. Although this book does not devote much discussion to the industry before 1860, my "Taming the Lightning: American Telegraphy as a Revolutionary Technology, 1832–1860" (Ph.D. diss., Case Western Reserve University, 1999) focused on that period. This series was important to that research because it gave valuable information on the history of many telegraph companies active in the 1840s and 1850s. Finally, the collection contains the records of Western Union's telephone and stock quotation subsidiaries, the American Speaking Telephone Company and Gold and Stock Telegraph Company.

Telegraph Technology

Good histories of telegraph technology include Paul Israel, *From Machine Shop to Industrial Laboratory: Telegraphy and the Changing Context of American Invention, 1830–1920* (Baltimore: Johns Hopkins University Press, 1992); Ken Beauchamp, *History of Telegraphy* (London: Institution of Electrical Engineers, 2001); Russell W. Burns, *Communications: An International History of the Formative Years* (London: Institution of Electrical Engineers, 2004); Lewis Coe, *The Telegraph: A History of Morse's Invention and Its Predecessors in the United States* (Jefferson, N.C.: Macfarland, 1993). Also useful are *The Papers of Thomas A. Edison*, vols. 1–5 (Baltimore: Johns Hopkins University Press, 1989–2004), and the Edison Papers Digital Edition at http://edison.rutgers.edu/digital.htm.

Valuable contemporary telegraph treatises include Alfred Vail, *The American Electro Magnetic Telegraph* (Philadelphia: Lea & Blanchard, 1845); Taliaferro P. Shaffner, *The Telegraph Manual: A Complete History and Description of the Semaphoric, Electric and Magnetic Telegraphs* (New York: Pudney and Russell, 1859); George B. Prescott, *History, Theory, and Practice of the Electric Telegraph* (Boston: Ticknor and Fields, 1860); Franklin L. Pope, *Modern Practice of the Electric Telegraph: A Handbook for Electricians and Operators* (New York: Russell Brothers, 1869); George B. Prescott, *The*

Speaking Telephone, Electric Light, and Other Recent Electrical Inventions (New York: D. Appleton, 1879); Thomas D. Lockwood, *Electricity, Magnetism, and Electric Telegraphy* (New York: D. Van Nostrand, 1883); William Maver Jr., *American Telegraphy and Encyclopedia of the Telegraph: Systems, Apparatus, Operation* (New York: Maver, 1909). Also of interest are two engineering reports. In 1867 Western Union hired British engineer Cromwell Fleet Varley to survey the condition of Western Union's lines and equipment. Varley's report is in the Western Union records. In 1877 the British Post Office sent two telegraph engineers, Henry C. Fischer and William H. Preece, to study the American telegraph industry in its entirety. Their "Joint Report upon the American Telegraph System" is in the British Post Office Archives, London. Paul Israel, director of the Thomas Edison Papers, allowed me to photocopy his copy, for which I am deeply grateful.

Telegraphy's Scientific Underpinnings

A good place to start on the relationship between advances in electrical science and early telegraph technology is my "Taming the Lightning: American Telegraphy as a Revolutionary Technology, 1832–1860," particularly chapters 1 and 2. Useful histories of science are J. L. Heilbron, *Electricity in the 17th and 18th Centuries: A Study in Early Modern Physics* (Berkeley: University of California Press, 1979); Edmund Whittaker, *A History of the Theories of Aether and Electricity*, vol. 1, *The Classical Theories* (New York: Philosophical Library, 1951); Robert V. Bruce, *The Launching of Modern American Science: 1846–1876* (New York: Alfred A. Knopf, 1987); P. M. Harman, *Energy, Force, and Matter: The Conceptual Development of Nineteenth-Century Physics* (Cambridge: Cambridge University Press, 1982); Michael Brian Schiffer, *Draw the Lightning Down: Benjamin Franklin and Electrical Technology in the Age of Enlightenment* (Berkeley: University of California Press, 2003). For an excellent account of amateur and public science in the era of telegraphy, consult Iwan Rhys Morus, *Frankenstein's Children: Electricity, Exhibition, and Experiment in Early Nineteenth Century London* (Princeton: Princeton University Press, 1998).

American physicist Joseph Henry played a central role in the theory and construction of electromagnets, a necessary foundation for Morse's telegraph. An excellent scientific biography of Henry is Albert E. Moyer, *Joseph Henry: The Rise of an American Scientist* (Washington, D.C.: Smithsonian Institution Press, 1997). A good summary of Henry's electromagnetic researches is Roger E. Sherman, "Joseph Henry's Contributions to the Electromagnet and the Electric Motor," *Rittenhouse* 12 (Oct. 1998): 97–106. The first eight volumes of the twelve-volume *Papers of Joseph Henry* (Washington, D.C.: Smithsonian Institution Press, 1972–2007) contain important material as well. Although Henry and Morse had an excellent professional and personal relationship from the late 1830s to mid-1840s, the lengthy litigation over Morse's telegraph patent led to a mutual enmity. Henry testified, albeit reluctantly, on behalf of Morse's opponents in several patent infringement cases. On their falling out, start with David Hochfelder, "Two Controversies in the Early History of Telegraphy," *IEEE Communications Magazine* 48 (Feb. 2010): 28–32.

The Telegraph Industry before 1866

In addition to Robert Luther Thompson's *Wiring a Continent*, economic historian Tomas W. Nonnenmacher has performed valuable research on the early industry. See his articles and dissertation: "State Promotion and Regulation of the Telegraph Industry, 1845–1860," *Journal of Economic History* 61 (Mar. 2001): 19–36; "Network Quality in the Early Telegraph Industry," *Research in Economic History* 23 (2005): 61–82; "Law, Emerging Technology and Market Structure: The Development of the Telegraph Industry: 1838–1868" (Ph.D. diss., University of Illinois at Urbana-Champaign, 1996).

Also valuable is James D. Reid, *The Telegraph in America: Its Founders, Promoters and Noted Men* (New York: Derby Brothers, 1879). Reid was a telegraph pioneer who began his career in 1845. By the time he wrote his book, he was a Western Union executive and editor of its in-house magazine, the *Journal of the Telegraph*. Because Reid was an industry insider, his book contains many personal anecdotes and biographical sketches, as well as information on the organization and operation of every telegraph company of any importance. The Western Union collection contains the records of most of the early telegraph companies as well. Also useful are Alexander Jones, *Historical Sketch of the Electric Telegraph, including Its Rise and Progress in the United States* (New York: George P. Putnam, 1852), and two short-lived telegraph journals, the *American Telegraph Magazine* published during 1852 and 1853 and *Shaffner's Telegraph Companion* published in 1854 and 1855.

For information on the early development and commercialization of the telegraph between 1832 and 1845, begin with biographies and personal papers of the principals. Morse is the subject of two excellent biographies. The most recent and definitive is Kenneth Silverman, *Lightning Man: The Accursed Life of Samuel F. B. Morse* (New York: Alfred A. Knopf, 2003). Somewhat dated but engagingly written is Carleton Mabee, *The American Leonardo: A Life of Samuel F. B. Morse*, rev. ed. (Fleischmanns, N.Y.: Purple Mountain Press, 2000). Morse's son Edward Lind Morse issued a two-volume collection of his father's papers, E. L. Morse, *Samuel F. B. Morse: His Letters and Journals* (New York: Houghton Mifflin, 1914). Morse's papers are at the Manuscript Division, Library of Congress. While biographers and historians have thoroughly mined them for his invention and early development of the telegraph, they are also useful for information on later developments, like the consolidations of the 1850s leading to the formation of the North American Telegraph Association in 1857. New York University also holds a small collection of Morse's papers. The Linda Hall Library in Kansas City, Missouri, has Morse's notebook from 1843 which contains details of his work on the pioneer Washington-to-Baltimore telegraph line.

The Western Union records at the National Museum of American History contain a small collection of letters between Morse and his business agent Amos Kendall. Kendall, who had been Jackson's postmaster general, is the subject of an excellent biography: Donald Cole, *A Jackson Man: Amos Kendall and the Rise of American Democracy* (Baton Rouge: Louisiana State University Press, 2004). Small collections of

Kendall's papers are at the Filson Historical Society, Louisville, Kentucky; Manuscript Division, Library of Congress; and Massachusetts Historical Society, Boston.

Alfred Vail was second in importance to Morse in the early development of the telegraph. Although there is no published biography of Vail, Stephen Ward Righter wrote a typescript biography in 1918. Commissioned by Vail's cousin and AT&T executive Theodore N. Vail, it is largely an attempt to enhance Vail's historical reputation at the expense of Morse's. Copies are at the New-York Historical Society, New Jersey Historical Society (Newark, N.J.), and the Local History Office of the Morristown (N.J.) Public Library. Vail's son also attempted to burnish his historical legacy by issuing an edited collection of his papers, J. Cummings Vail, *Early History of the Electro-Magnetic Telegraph from Letters and Journals of Alfred Vail* (New York: Hine Brothers, 1914). Vail's personal papers are at two locations. The Smithsonian Institution Archives holds the Vail Telegraph Collection, which is invaluable for material on the invention and early development of the telegraph. A smaller collection at the New Jersey Historical Society in Newark contains his diaries between 1850 and his death in 1858, as well as some family correspondence.

Telegraph pioneer Ezra Cornell is the subject of a readable biography, though it lacks notes: Philip Dorf, *The Builder: A Biography of Ezra Cornell* (New York: Macmillan, 1952). Cornell's personal papers are quite extensive and are held at Cornell University. They are also viewable online at http://collections.library.cornell.edu/ezra/ (accessed 10 Aug. 2011).

Henry O'Rielly, initially a Morse ally and later a rival, left an enormous collection of his papers to the New-York Historical Society. In many ways this collection serves as a counterweight to the Morse and Vail papers, because it illuminates the activities of companies not subject to Morse's control. Because O'Rielly challenged the validity of Morse's patent, resulting in many years of litigation, his papers give details of rival telegraph designs such as Alexander Bain's chemical telegraph and Royal E. House's Roman letter printer. A smaller collection of O'Rielly's papers is at the Rochester (N.Y.) Public Library. Salmon P. Chase acted as O'Rielly's lawyer and business adviser during the late 1840s and early 1850s, and relevant correspondence is in Chase's papers at the Manuscript Division, Library of Congress. Chase's diaries are reproduced in John Niven, ed., *The Salmon P. Chase Papers*, vol. 1, *Journals, 1829–1872* (Kent, Ohio: Kent State University Press, 1993).

Samuel Morse gave Maine congressman Francis O. J. Smith a share of his patent in 1838 to help him get federal funding and to convince Congress to buy his patent. Smith, a difficult and litigious man, turned out to be an impediment to the development of the industry until the American Telegraph Company bought his telegraph holdings in the late 1850s. Smith is the subject of a biography: Thomas L. Gaffney, "Maine's Mr. Smith: A Study of the Career of Francis O. J. Smith, Politician and Entrepreneur" (Ph.D. diss., University of Maine, 1979). His personal papers are at the Maine Historical Society, Portland.

Western Union's first two presidents, Hiram Sibley (1856–66) and Jeptha Wade (1866–67), left useful collections of papers, Sibley's at University of Rochester (N.Y.)

and Wade's at the Western Reserve Historical Society, Cleveland, Ohio. In addition to giving insight on the early history of Western Union, these two collections also contain material on Western Union's line to the Pacific and its abortive attempt to link the continents via its Russian extension line.

The John Dean Caton Papers at the Manuscript Division, Library of Congress, trace the history of one of the six signatories to the North American Telegraph Association contract. Caton was president of the Illinois and Mississippi Telegraph Company, a small regional telegraph company operating in Illinois, Iowa, and Missouri. His papers give details on how early companies operated, on the effects of the Civil War on the industry, and on the great telegraph mergers of 1866.

Material related to the formation and operations of the American Telegraph Company between its incorporation in 1856 and its absorption by Western Union in 1866 is in the Abram Hewitt Papers, formerly housed at the Passaic County Historical Society, Passaic, New Jersey, now at the Cooper Union, New York. The Peter Cooper Manuscripts at the Cooper Union are also useful. These collections also contain information on the several attempts to lay an Atlantic telegraph cable, as does the Cyrus Field Papers, New York Public Library.

The Telegraph during the Civil War

The best history of the Signal Corps from its 1860 formation onward is Rebecca Robbins Raines, *Getting the Message Through: A Branch History of the U.S. Army Signal Corps* (Washington, D.C.: Center of Military History, United States Army, 1996). J. Willard Brown, *The Signal Corps U.S.A. in the War of the Rebellion* (Boston: U.S. Veteran Signal Corps Association, 1896) contains a wealth of detail, personal anecdotes, and biographical sketches. The best treatment of Chief Signal Officer Albert Myer is Paul Joseph Schieps, "Albert James Myer, Founder of the Army Signal Corps: A Biographical Study" (Ph.D. diss., American University, 1966). Two collections of Myer's papers are at the U.S. Army Military History Institute, Carlisle Barracks, Pennsylvania, and the Manuscript Division, Library of Congress. The Library of Congress has the former collection on microfilm as well. Material on the relationship between the Signal Corps and U.S. Military Telegraph, and on the Beardslee field telegraphs, is in the National Archives, RG 111, Records of the Office of the Chief Signal Officer, particularly entries 1 and 4-A, which contain outgoing letters and received letters and telegrams.

The best place to start researching the U.S. Military Telegraph (USMT) is William R. Plum's two-volume *Military Telegraph during the Civil War in the United States* (Chicago: Jansen, McClurg, 1882). Plum was a military telegrapher who collected diaries, letters, and reminiscences to write the volumes. Plum was also active in organizing the Society of the United States Military Telegraph Corps and its annual reunions from 1882 to 1900. Proceedings of the reunions are at the New York Public Library. A readable popular history is Tom Wheeler, *Mr. Lincoln's T-Mails: The Untold Story of How Abraham Lincoln Used the Telegraph to Win the Civil War* (New York: HarperCollins, 2006).

A good supplement to Plum's book is the War Department's massive published collection of documents, *The War of the Rebellion: A Compilation of the Official Records of the Union and Confederate Armies.* Fortunately, the *Official Records* are now online and keyword-searchable through Cornell University's Making of America Web site, http://digital.library.cornell.edu/m/moawar/waro.html (accessed 7 Apr. 2012). Material in National Archives, RG 107, Records of the Office of the Secretary of War, Telegrams Collected by the Office of the Secretary of War, contains information on War Department supervision of the telegraphs generally and some material on the censorship of telegraphic news. They are available as National Archives Microform Publication M504.

Roscoe Pound relied heavily on the *Official Records* to write a jaundiced article about the conduct of USMT operators, "The Military Telegraph in the Civil War," *Proceedings of the Massachusetts Historical Society* 66 (1942): 185–203. Letters and diaries left by military operators show instead that USMT operators generally conducted themselves honorably and courageously. I have relied on six manuscript collections, in addition to Plum's book, to research the wartime experiences of USMT personnel. The most extensive and useful is the William L. Gross Papers at the Western Reserve Historical Society, Cleveland, Ohio. Captain (later breveted to Colonel) Gross was assistant superintendent of both the USMT and the South Western Telegraph Company, and his papers give an excellent view into the relationship between the commercial telegraph companies and the USMT, as well as details of army life and daily USMT operations. The other five collections are at the Manuscript Division, Library of Congress. Robert C. Clowry was at the same rank as Gross, and he left a small collection of papers giving some information on his wartime experiences. Although George Kennan was not a USMT operator, his papers detail the wartime experiences of commercial operators and give insight into the career and character of Western Union and USMT superintendent Anson Stager. David Homer Bates, stationed at the War Department telegraph office, left a diary and some correspondence that illuminates life in wartime Washington. Donald E. Markle published Bates's diary and his annotations are of value: Markle, *The Telegraph Goes to War: The Personal Diary of David Homer Bates, Lincoln's Telegraph Operator* (Lynchburg, Va.: Schroeder Publications, 2005). Bates wrote a volume of his wartime reminiscences as well: *Lincoln in the Telegraph Office: Recollections of the United States Military Telegraph Corps during the Civil War* (New York: Century, 1907). For the experiences of USMT operators in the field, Luther Rose's detailed diary is probably the best source. Rose was a telegrapher in the Army of the Potomac and his diary covers from April 1863 to October 1864. Joseph Schnell was also an operator for the Army of the Potomac, and his small collection of papers is of some use.

Thompson's *Wiring a Continent* covers well the wartime history of the industry itself. To supplement his coverage, consult the Hiram Sibley and John Dean Caton Papers. Also useful are the Norvin Green Family Papers, University of Kentucky, Lexington. Proceedings of the annual meetings of the North American Telegraph Association are extant, and they contain relevant material on the relationship among the companies and between the companies and the War Department.

The history of the telegraph in the Confederacy is considerably harder to research. A good overview is J. Cutler Andrews, "The Southern Telegraph Company, 1861–1865: A Chapter in the History of Wartime Communications," *Journal of Southern History* 30 (Aug. 1964): 319–44. Although records are incomplete, there are a few manuscript sources yielding information on the Confederate telegraph system. The largest, and the collection that Andrews relied on for his 1964 article, is the Southern Telegraph Company Papers at the Carnegie Library, Pittsburgh. Smaller collections include the Morris Family Collection, Museum of the Confederacy, Richmond; Business Records, Confederate States of America Post Office, Accession 22724, Library of Virginia, Richmond; Thomas Hicks Wynne Papers, part of Robert Alonzo Brock Collection, microfilm held at Library of Virginia, Richmond; War Department Collection of Confederate Records, Confederate Military Telegraph Accounts, RG 109, National Archives. Some telegraph material is in Theron Wierenga, ed., *Official Documents of the Post Office Department of the Confederate States of America* (Holland, Mich.: Theron Wierenga, 1979). For an account of the internecine conflict in the Southern telegraph industry, see C. P. Culver, "The Southern Telegraph," 25 Jan. 1863, Confederate Imprints, University of Georgia Library. Also of interest are Edward Rosewater's diaries in the Rosewater Family Papers, American Jewish Archives, Cincinnati, Ohio. Rosewater was a telegraph operator working in the South when the war began. He returned north in the summer of 1861 and spent time in the War Department telegraph office. Some material related to the southern telegraph industry is in the Samuel F. B. Morse Papers, Manuscript Division, Library of Congress.

The Telegraph Industry between 1866 and 1909

The best source for the history of telegraphy in this period is the Western Union corporate archive held at the National Museum of American History. The letterbooks of William Orton and Norvin Green are especially useful, as are the records of telegraph companies that competed with Western Union before that company absorbed them. National Archives, RG 27, Records of the Weather Bureau, entries 6, 8, 12, 21, 22, and 23, contains some Signal Service correspondence on Western Union's competitors in the early 1870s.

In addition to the Western Union and Signal Service records, several collections of personal papers are valuable. Although William Orton left no personal papers, Norvin Green did. Some of it survives in the George Douglass Papers, University of Kentucky, Lexington; the Norvin Green Collection, Filson Historical Society, Louisville, Kentucky; and Norvin Green Family Papers, University of Kentucky. The papers of Western Union and Gold and Stock Telegraph Company executive Marshall Lefferts are at the New-York Historical Society. They contain some information on Western Union and Gold and Stock operations in the 1870s, but the bulk relates to the efforts of Daniel Craig and Thomas Edison to develop a rapid automatic telegraph. The papers of Edwin D. Morgan, a director of Western Union and prominent New York politician, contain correspondence to and from William Orton, William H. Vanderbilt, Jay Gould, and other men prominent in the telegraph industry. Morgan's papers are at the New York State Library Manuscripts Room, Albany. On Jay Gould's telegraph activities, start with the two standard biographies, Julius Grodinsky, *Jay Gould: His Business Career, 1867–1892* (Philadelphia: University of Pennsylvania Press, 1957); and Maury Klein, *The Life and Legend of Jay Gould* (Baltimore: Johns Hopkins University Press, 1986). Gould left no personal papers, but some of his correspondence related to telegraph matters is in the papers of New Hampshire politician William E. Chandler at the Manuscript Division, Library of Congress.

The operators' journal the *Telegrapher*, published between 1864 and 1877, contains technical correspondence, news of the profession and industry, and coverage of the postal telegraph debate. Telegrapher William Andrews Manning kept a diary between 1867 and 1872 that shows what working conditions were like for operators in that period. Manning was also a local leader of the Telegraphers' Protective League and was active in an 1870 operators' strike. His diary is at the Western Reserve Historical Society, Cleveland, Ohio.

Postal Telegraph Movement

This is perhaps the most thoroughly studied aspect of the history of American telegraphy. John's *Network Nation*, chapters 3 to 5, covers the issue completely. A useful shorter treatment is Menahem Blondheim, "Rehearsal for Media Regulation: Congress versus the Telegraph-News Monopoly, 1866–1900," *Federal Communications Law Journal* 56 (2004): 299–328. A study of the influence of the British nationalization on the American postal telegraph movement is David Hochfelder, "A Comparison of the Postal Telegraph Movement in Great Britain and the United States, 1866–1900," *Enterprise and Society* 1 (Dec. 2000): 739–61. An excellent discussion of the relationship between Western Union and the antimonopoly movement is Joshua D. Wolff, "'The Great Monopoly': Western Union and the American Telegraph, 1845–1893" (Ph.D. diss., Columbia University, 2008). Lester Lindley's 1971 dissertation, republished without revision by Arno Press in 1975, has utility for locating primary sources like government documents, but exercise caution regarding matters of interpretation: Lester G. Lindley, *The Constitution Faces Technology: The Relationship of the National Government to the Telegraph, 1866–1884* (New York: Arno Press, 1975).

A key reason for the postal telegraph movement being so thoroughly studied is that the major primary sources are government documents, particularly congressional hearings, committee reports, and proceedings of floor debates from the *Congressional Globe* and *Congressional Record*. These are voluminous, especially during the 41st to 43d Congresses (1869–75). The letterbooks of Western Union presidents William Orton and Norvin Green contain much useful information as well.

The personal papers of key participants contain material related to the postal telegraph. Particularly helpful at the Library of Congress's Manuscript Division are the papers of John Sherman, James Garfield, John A. J. Creswell, and Gardiner Hubbard. Sherman was partially responsible for the National Telegraph Act of 1866. Garfield chaired the House Appropriations Committee at a crucial period of the movement. Alongside his manuscript papers, a published edition of his diary contains useful annotation: Harry James Brown and Frederick D. Williams, eds., *The Diary of James A. Garfield*, 4 vols. (East Lansing: Michigan State University Press, 1967–81). As President Grant's postmaster general, Creswell was a strong advocate of nationalizing the telegraph under the terms of the 1866 act. The Hubbard Family Papers are especially useful because of Gardiner Hubbard's long and persistent advocacy of postal telegraphy between 1868 and the early 1890s.

On the Wilson administration's takeover of the telegraphs and telephones during World War 1, start with Jonathan Reed Winkler, *Nexus: Strategic Communications and American Security during World War I* (Cambridge: Harvard University Press, 2008). Arthur S. Link's complete published edition of Woodrow Wilson's papers are also indispensable. Useful information on the views of members of Wilson's cabinet is in E. David Cronon, ed., *The Cabinet Diaries of Josephus Daniels, 1913–1921* (Lincoln: University of Nebraska Press, 1963). Some of the papers of Albert Sidney Burleson, Wilson's postmaster general, are at the Manuscript Division, Library of Congress. National Archives, RG 28, Records of the Post Office Department, especially entries 38 and 47, give details on the Post Office's wartime operation of the wires. Also consult the voluminous 1919 hearings on the issue, "Government Control of the Telegraph and Telephone Systems: Hearings on H.J. Res. 368, House Committee on Post Offices and Post Roads," 65th Cong., 3d sess., 1919.

The Army Signal Service and Weather Reporting

In addition to Rebecca Raines's branch history of the Signal Corps, James Fleming's work is a good place to start on the army's weather reporting activities: James Rodger Fleming, *Meteorology in America, 1800–1870* (Baltimore: Johns Hopkins University Press, 1990); and Fleming, "Storms, Strikes, and Surveillance: The U.S. Army Signal Office, 1861–1891," *Historical Studies in the Physical and Biological Sciences* 30 (2000): 315–32. See also Donald R. Whitnah, *A History of the United States Weather Bureau* (Urbana: University of Illinois Press, 1965); and Joseph M. Hawes, "The Signal Corps and Its Weather Service, 1870–1890," *Military Affairs* 30 (Summer 1966): 68–76. Also quite useful are the *Annual Reports of the Chief Signal Officer* between 1870 and the early 1890s. Economic historian Erik D. Craft found that the social savings of the army's weather forecasts far outweighed their costs, particularly with regard to the prevention of shipwrecks: Craft, "The Value of Weather Information Services for Nineteenth-Century Great Lakes Shipping," *American Economic Review* 88 (1998): 1059–76.

On the fraught relationship between Western Union and the Army Signal Service, start with this report, "Signal-Service and Telegraph Companies," 42d Cong., 2d Sess., H.R. Rep. 69, 9 May 1872. William Orton's letterbooks contain correspondence on the issue as well. National Archives, RG 27, Records of the Weather Bureau, has a wealth of information on the relationship between the Signal Service, Western Union, and Western Union's competitors. Entries 6, 8, 12, 21, 22, and 23 were all useful.

On the role of the Signal Service during the 1877 Great Strike, turn to Robert V. Bruce, *1877: Year of Violence* (Chicago: Ivan R. Dee, 1987 [1959]). The Rutherford B.

Hayes Presidential Library, Fremont, Ohio, holds copies of the Signal Service telegrams that Hayes received.

The Signal Service turned over its weather-reporting function to the U.S. Department of Agriculture in the early 1890s and thereafter devoted itself to military signaling. For a lively account of its role in the Spanish-American War, see Howard A. Giddings, *Exploits of the Signal Corps in the War with Spain* (Kansas City: Hudson-Kimberly Publishing Co., 1900).

The Telegraph and the Newspaper Press

Helpful histories of telegraphic newsgathering are Frederic Hudson, *Journalism in the United States, from 1690 to 1872* (New York: Harper & Brothers, 1873); Victor Rosewater, *History of Cooperative News-Gathering* (New York: D. Appleton, 1930); Michael Schudson, *Discovering the News: A Social History of American Newspapers* (New York: Basic Books, 1978); Dan Schiller, *Objectivity and the News: The Public and the Rise of Commercial Journalism* (Philadelphia: University of Pennsylvania Press, 1981); Richard A. Schwartzlose, *The Nation's Newsbrokers*, 2 vols. (Evanston, Ill.: Northwestern University Press, 1989 and 1990); Menahem Blondheim, *News over the Wires: The Telegraph and the Flow of Public Information in America, 1844–1897* (Cambridge: Harvard University Press, 1994); David Mindich, *Just the Facts: How "Objectivity" Came to Define American Journalism* (New York: New York University, 1998); Richard L. Kaplan, *Politics and the American Press: The Rise of Objectivity, 1865–1920* (New York: Cambridge University Press, 2002).

On the formation and early history of the New York Associated Press, start with Alexander Jones, *Historical Sketch of the Electric Telegraph*. The Francis O. J. Smith Papers at the Maine Historical Society, Portland, contain correspondence related to Smith's conflict with the New York Associated Press over telegraph rates and terms of service in the late 1840s and early 1850s.

On Civil War reporting, start with J. Cutler Andrews's exhaustive two-volume history of Northern and Southern war reporting: *The North Reports the Civil War* (Princeton: Princeton University Press, 1955) and *The South Reports the Civil War* (Princeton: Princeton University Press, 1970). Another readable account is Louis M. Starr, *Bohemian Brigade: Civil War Newsmen in Action* (New York: Alfred A. Knopf, 1954). On wartime censorship of the telegraph, see Richard B. Kielbowicz, "The Telegraph, Censorship, and Politics at the Outset of the Civil War," *Civil War History* 15 (1994): 95–118; and Menahem Blondheim, " 'Public Sentiment Is Everything': The Union's Public Communications Strategy and the Bogus Proclamation of 1864," *Journal of American History* 89 (2002): 869–99. Kielbowicz drew heavily on unpublished 1862 congressional hearings on telegraphic censorship, *Allegations of Government Censorship of Telegraphic News Reports*, House Committee on the Judiciary, 37th Cong., 2d sess., HJ-T.1 12. The committee's report on these hearings is House Report 64, 37th Cong., 2d sess., 20 March 1862. Some correspondence related to telegraphic censorship is in the William H. Seward Papers, Manuscript Division, Library of Congress. The memoirs of the Associated Press Washington correspondent

Lawrence Gobright, *Recollection of Men and Things at Washington, during the Third of a Century* (Philadelphia: Claxton, Remsen & Haffelfinger, 1869), contain information on press censorship as well. Many of Gobright's dispatches are retained in National Archives, RG 107, Office of the Secretary of War, Telegrams Sent by the Field Office of the Military Telegraph.

On relations between Western Union, the New York Associated Press, and regional press associations after the Civil War, see Peter R. Knights, *The Press Association War of 1866–1867* (Austin: Association for Education in Journalism, 1968); and Alex Nalbach, "'Poisoned at the Source'? Telegraphic News Services and Big Business in the Nineteenth Century," *Business History Review* 77 (Winter 2003): 577–610. The published proceedings of the Western Associated Press are extant between 1867 and 1888. For a view into Western Union's relationship with press associations, consult unpublished House Judiciary Committee hearings from 1875: "In the Matter of the Western Union Telegraph Co.," 43d Cong., 2d sess., HJ-T.2. The personal papers of newspaper publishers and newsbrokers Murat Halstead (Cincinnati Historical Society), William Henry Smith (Indiana Historical Society, Indianapolis), and Manton Marble (Manuscript Division, Library of Congress) are also useful.

On the telegraph's relationship to the changing psychology of news consumption and to prose composition, my thinking has benefited from the work of literary scholar Richard Menke, particularly "Telegraphic Realism: Henry James's *In the Cage*," *PMLA* 115 (2000): 975–90; "Media in America, 1881: Garfield, Guiteau, Bell, Whitman," *Critical Inquiry* 31 (2005): 638–64; and *Telegraphic Realism: Victorian Fiction and Other Information Systems* (Stanford, Calif: Stanford University Press, 2008).

The Telegraph and Financial Markets

Useful general histories of financial markets include Ranald C. Mitchie, *The Global Securities Market: A History* (Oxford: Oxford University Press, 2006); Lawrence E. Mitchell, *The Speculation Economy: How Finance Triumphed over Industry* (San Francisco: Bennett-Koehler Publishers, 2007); and Steve Fraser, *Every Man a Speculator: A History of Wall Street in American Life* (New York: HarperCollins, 2005). An interesting account of technology and commodity trading in the twenty-first century is Caitlin Zaloom, *Out of the Pits: Traders and Technology from Chicago to London* (Chicago: University of Chicago Press, 2006).

Economic historians have done the most to investigate the relationship between communication technology and the structure and operations of financial markets. Begin with Kenneth Garbarde and William L. Silber, "Technology, Communication, and the Performance of Financial Markets," *Journal of Finance* 33 (1978): 819–32. Richard B. DuBoff and Alexander Field more fully explore the role of the telegraph in American financial markets: DuBoff, "Business Demand and the Development of the Telegraph in the United States, 1844–1860," *Business History Review* 54 (1980): 459–79; DuBoff, "The Telegraph and the Structure of Markets in the United States, 1845–1890," *Research in Economic History* 8 (1983): 253–77; Field, "The Magnetic Telegraph, Price and Quantity Data, and the New Management of Capital," *Journal of Economic*

History 52 (1992): 401–13; Field, "The Telegraphic Transmission of Financial Asset Prices and Orders to Trade: Implications for Economic Growth, Trading Volume and Securities Market Regulation," *Research in Economic History* 18 (1998): 145–84.

The work of economic sociologist Alex Preda has especially stimulated my thinking. One of Preda's key insights was to ask why the stock ticker was so important to the rise of modern financial markets. See Preda, "Socio-Technical Agency in Financial Markets: The Case of the Stock Ticker," *Social Studies of Science* 36 (2006): 753–82; and Preda, *Framing Finance: The Boundaries of Markets and Modern Capitalism* (Chicago: University of Chicago Press, 2009).

Although it is risky to generalize from the experiences of a few individuals, accounts left by stock traders can provide a window into the mechanics and psychology of late nineteenth-century investing and trading. Edward Neufville Tailer's detailed and extensive diaries at the New-York Historical Society offer an unmediated view of stock trading in the 1870s and 1880s. See also Henry Clews, *Twenty-Eight Years in Wall Street* (New York: Irving, 1888); and Edmund Clarence Stedman, "Life 'On the Floor': The New York Stock Exchange from Within," *Century Magazine* 68 (Nov. 1903): 1–20. On Jesse Livermore's experiences as a bucket shop patron and stock trader, see Richard Smitten, *Jesse Livermore: World's Greatest Stock Trader* (New York: John Wiley and Sons, 2001); and Edwin Lefevre, *Reminiscences of a Stock Operator* (New York: George H. Doran, 1923). Richard D. Wyckoff's reminiscences cover a later period but include information about bucket shops as well as the mechanics of stock trading: Wyckoff, *Wall Street Ventures & Adventures through Forty Years* (Greenville, S.C.: Traders Press, 1986 [1930]).

Robert Sobel is the leading historian of New York's stock markets, particularly the New York Stock Exchange and the American Stock Exchange. See his books: *The Big Board: A History of the New York Stock Market* (New York: Free Press, 1965); *The Curbstone Brokers: The Origins of the American Stock Exchange* (New York: Macmillan, 1970); and *Inside Wall Street* (New York: W. W. Norton, 1977). The records of the New York Stock Exchange remain at the exchange and are open to serious researchers.

The best history of the Chicago Board of Trade is Jonathan Lurie, *The Chicago Board of Trade, 1859–1905: The Dynamics of Self-Regulation* (Urbana: University of Illinois Press, 1979). Supplement Lurie's book with the official and quite detailed history of the board, Charles H. Taylor, *History of the Board of Trade of the City of Chicago* (Chicago: Robert O. Law, 1917). An excellent description of the role of the Chicago Board of Trade in the economy of the Midwest is in William Cronon, *Nature's Metropolis: Chicago and the Great West* (New York: W. W. Norton, 1991). Cronon's descriptions of futures trading and corners are unsurpassed. For a fictional account that fits well with Cronon's book, see Frank Norris, *The Pit: A Story of Chicago* (New York: Doubleday, Page, 1903). The Chicago Board of Trade records are at University of Illinois, Chicago, and researchers must get board permission to gain access to them.

For the activities of smaller regional exchanges, Bradford Scharlott's work on the Cincinnati Merchants' Exchange has been the most extensive. See his detailed dissertation, "The Telegraph and the Integration of the U.S. Economy: The Impact of Electrical Communications on Interregional Prices and the Commercial Life of

Cincinnati" (Ph.D. diss., University of Wisconsin, 1986); and a more recent article, "Communication Technology Transforms the Marketplace: The Effect of the Telegraph, Telephone, and Ticker on the Cincinnati Merchants' Exchange," *Ohio History* 113 (2004): 4–17. Supplement Scharlott's work with the records of the Cincinnati Chamber of Commerce and Merchants' Exchange held at the Cincinnati Historical Society. The New York Public Library has annual reports of the New York Produce Exchange from 1879 onward and of the Consolidated Stock Exchange from 1886 until its demise in the 1920s. These contain information on ticker service and their relationship with the major exchanges in their fields, the Chicago Board of Trade and New York Stock Exchange respectively.

On bucket shops and the cultural and economic confusion between speculation and gambling, start with David Hochfelder, "'Where the Common People Could Speculate': The Ticker, Bucket Shops, and the Origins of Popular Participation in Financial Markets," *Journal of American History* 93 (Sept. 2006): 335–58. Also valuable are Jonathan Ira Levy, "Contemplating Delivery: Futures Trading and the Problem of Commodity Exchange in the United States, 1875–1905," *American Historical Review* 111 (June 2006): 307–35; Ann Fabian, *Card Sharps, Dream Books, and Bucket Shops: Gambling in Nineteenth-Century America* (Ithaca, N.Y.: Cornell University Press, 1990); Cedric B. Cowing, *Populists, Plungers, and Progressives: A Social History of Stock and Commodity Speculation, 1890–1936* (Princeton: Princeton University Press, 1965). For a contemporary view of bucket shopping, see John Hill Jr., *Gold Bricks of Speculation: A Study of Speculation and Its Counterfeits, and an Exposé of the Methods of Bucketshop and "Get-Rich-Quick" Swindles* (New York: Arno Press, 1975; reprint of Chicago: Lincoln Book Concern, 1904). Hill was a Chicago Board of Trade investigator who led efforts to stamp out the bucket shops, and his book gives details of how bucket shops operated.

The Telegraph Industry after 1909

In addition to the Western Union and AT&T archives, the best sources for information on the telegraph industry in the twentieth century are articles from business periodicals and government documents, particularly congressional hearings and reports and FCC investigations. Other useful sources are Hyman Howard Goldin, "The Domestic Telegraph Industry and the Public Interest: A Study in Public Utility Regulation" (Ph.D. diss., Harvard University, 1950); and the President's Communications Policy Board, *Telecommunications: A Program for Progress* (Washington, D.C.: GPO, 1951). This report is reprinted in John M. Kittross, ed., *Documents in American Telecommunications Policy*, vol. 2 (New York: Arno Press, 1977). Western Union published an engineering journal, the *Western Union Technical Review*, from 1947 to 1969. Although inferior in quality to the highly regarded *Bell System Technical Journal*, it offers a view into Western Union's belated attempt to build an internal research-and-development capability. Retired Western Union executive George Oslin wrote two autobiographical accounts of his service that provide a personal window into the company's operations: *The Story of Telecommunications* (Macon, Ga.: Mercer University Press, 1992) and *One Man's Century: From the Deep South to the Top of the Big Apple* (Macon, Ga.: Mercer University Press, 1998).

The Telephone and Telegraph

The telephone is one of the best-studied technologies in American history, with a large literature on its technological and organizational development. An important reason for this is that AT&T maintained a commitment to its history, supporting an extensive corporate archive open to researchers and commissioning a telephone history series through Johns Hopkins University Press. The AT&T Archives are located in Warren, New Jersey, and they contain extensive records of AT&T and its predecessors and subsidiaries, including the precursor national Bell companies; Bell's manufacturing arm, Western Electric; and its research-and-development arm, Bell Telephone Laboratories. That archive is the starting place for any research on the history of American telephony. The "old" AT&T merged with regional operating company SBC in 2005, with the merged company adopting the AT&T name. The SBC Archives in San Antonio, Texas, focus more on the local and regional Bell operating companies.

The Johns Hopkins/AT&T Series in Telephone History published four very useful studies in the 1980s: Neil Wasserman, *From Invention to Innovation: Long Distance Telephone Transmission at the Turn of the Century* (Baltimore: Johns Hopkins University Press, 1985); Robert W. Garnet, *The Telephone Enterprise: The Evolution of the Bell System's Horizontal Structure, 1876–1909* (Baltimore: Johns Hopkins University Press, 1985); Kenneth Lipartito, *The Bell System and Regional Business: The Telephone in the South, 1877–1920* (Baltimore: Johns Hopkins University Press, 1989); and George David Smith, *The Anatomy of a Business Strategy: Bell, Western Electric, and the Origins of the American Telephone Industry* (Baltimore: Johns Hopkins University Press, 1985). An excellent history of Western Electric is Stephen B. Adams and Orville R. Butler, *Manufacturing the Future: A History of Western Electric* (Cambridge: Cambridge University Press, 1999).

On the invention and early development of the telephone, start with Rosario Joseph Tosiello, *The Birth and Early Years of the Bell Telephone System, 1876–1880* (New York: Arno Press, 1979). These articles are also useful: David A. Hounshell, "Elisha Gray and the Telephone: On the Disadvantages of Being an Expert," *Technology and Culture* 16 (1975): 133–61; W. Bernard Carlson, "Entrepreneurship in the Early Development of the Telephone: How Did William Orton and Gardiner Hubbard Conceptualize the New Technology?" *Business and Economic History* 23 (1994): 161–92; Bernard S. Finn, "Bell and Gray: Just a Coincidence?" *Technology and Culture* 50 (2009): 193–201; Christopher Beauchamp, "Who Invented the Telephone? Lawyers, Patents, and the Judgments of History," *Technology and Culture* 51 (Oct. 2010): 854–78. An excellent contemporary account of telephone technology is George Prescott, *Bell's Electric Speaking Telephone: Its Invention, Construction, Application, Modification and History* (New York: D. Appleton, 1884). J. Leigh Walsh, *Connecticut Pioneers in Telephony: The Origin and Growth of the Telephone Industry in Connecticut* (New Haven,

Conn.: Telephone Pioneers of America, 1950) contains useful material on a potential
Jay Gould and Bell alliance in 1879 that partially prompted Western Union to exit the
telephone industry.

The extensive personal papers of Alexander Graham Bell and his father-in-law
(and perennial postal telegraph advocate) Gardiner Hubbard are both at the Manuscript Division, Library of Congress. These collections illuminate Bell's work on harmonic telegraph systems as well as his work on telephony. The best biography of Bell
is Robert V. Bruce, *Alexander Graham Bell and the Conquest of Solitude* (Boston:
Little, Brown, 1973). Some of Elisha Gray's papers are at the National Museum of
American History's Archives Center. Also useful is Elisha Gray, *Experimental Researches in Electro-Harmonic Telegraphy and Telephony, 1867–1878* (New York: Russell
Brothers, Printers, 1878).

On the Bell System's world-class research and development operations, start with
A History of Engineering and Science in the Bell System, 7 vols. (New York: Bell Telephone Laboratories, 1975–85) particularly the first volume covering the period between 1875 and 1925. Because these volumes are quite technical, supplement them
with these articles: John V. Langdale, "The Growth of Long-Distance Telephony
in the Bell System, 1875–1907," *Journal of Historical Geography* 4:2 (1978): 143–59;
Lillian Hoddeson, "The Emergence of Basic Research in the Bell Telephone System,
1875–1915," *Technology and Culture* 22 (1981): 512–44; Milton Mueller, "The Switchboard Problem: Scale, Signaling, and Organization in Manual Telephone Switching,
1877–1897," *Technology and Culture* 30 (1989): 534–60; Venus Green, "Goodbye Central: Automation and the Decline of 'Personal Service' in the Bell System, 1878–1921,"
Technology and Culture 36:4 (Oct. 1995): 912–49.

Theodore N. Vail left no extant personal papers, although much of his official correspondence survives in the AT&T Archives. He is the subject of a dated and uncritical biography, Albert Bigelow Paine, *In One Man's Life: Being Chapters from the Personal and Business Career of Theodore N. Vail* (New York: Harper Brothers, 1921). Vail
commissioned a collection of his public statements: Theodore Newton Vail, *Views
on Public Questions: A Collection of Papers and Addresses of Theodore Newton Vail,
1907–1917* (privately printed, 1917). A good summary of his career at AT&T is Louis
Galambos, "Theodore N. Vail and the Role of Innovation in the Modern Bell System,"
Business History Review 66 (Spring 1992): 95–126. A sympathetic account of Vail's
corporate progressivism is Alan Stone, *Public Service Liberalism: Telecommunications
and Transitions in Public Policy* (Princeton: Princeton University Press, 1991). More
critical is Noobar Danielian, *AT&T: The Story of Industrial Conquest* (New York: Vanguard Press, 1939); and Federal Communications Commission, *Investigation of the
Telephone Industry in the United States* (Washington, D.C.: GPO, 1939), 76th Cong.,
1st sess., H.R. Doc. 340, 14 June 1939. Vail's predecessor William Forbes is the subject
of an uncritical and unsourced biography, Arthur S. Pier, *Forbes: Telephone Pioneer*
(New York: Dodd, Mead, 1953).

Tuley, Murray, 131

Tumulty, Joseph, 67, 69

Twain, Mark (Samuel Clemens), 97

United Press, 86

United States Postal Telegraph Co., 59, 62

United States Telegraph Co., 29–31, 37–38, 55

U.S. Dept. of Justice, 134, 139, 159, 163, 168

U.S. Military Telegraph, 6–7, 10–28, 80, 90, 104, 180

Vail, Alfred, 75, 83, 87, 91

Vail, Theodore N., 78, 139, 155, 158–64

Van Antwerp, William C., 127, 134, 136

Van Duzer, John, 6, 17–21, 26–27

Van Hoevenburgh, Henry, 106–7

Van Horne, John, 45–46

Van Riper, Lewis C., 121

Van Rysselberghe, François, 156

Vanderbilt, Cornelius, 31, 39, 63, 120, 173–74

Vanderbilt, William H., 39, 47, 50, 149, 151–52, 173–74

Vilas, William, 77

Wade, Jeptha, 29–31, 55

Wanamaker, John, 45–46, 65, 67

Ward, George G., 81

Washburn, Cadwallader, 58, 62–63, 178

Washburne, Elihu, 45, 52, 62, 178

Waterbury, John I., 159

Watson, Thomas A., 145–46

Webster, Daniel, 97

Welles, Gideon, 15

Western Associated Press, 74, 82, 85–86, 91

Western Electric Co., 153, 164

Western Union Telegraph Co.: acquisition by AT&T, 5, 77–78, 139, 158–64; acquisition of Postal, 168–69; and antimonopoly sentiment, 44–45, 64–66, 70–71; and bucket shops, 122, 127–34; during Civil War, 7–9, 18, 24–28; Commercial News Dept., 107–9; competitive strategy of, 37–43, 177; line to Pacific, 28–29; lobbying activities of, 45–47; managers of, 18–19; and mergers of 1866, 3–4, 6–7, 30–33, 36; message length statistics, 79; and postal telegraph, 34–36, 43–67, 70; postwar decline of, 169–75; railroad rights-of-way, 35, 39–40, 58, 160–61; reduced-rate deferred telegrams, 76–78, 80–81, 162; relations with press, 40–44, 74, 83–86, 91, 96, 164, 180; research and development, 165–72; Russian-American Telegraph, 28–29; and Signal Service weather reports, 16, 59–62; and sports gambling, 113–15; telegraphers' strikes, 23; and telegraph innovation, 41–43, 102, 140–44, 173–74, 177; and telephone, 138–39, 144–57; ticker business, 106–15, 119; valuation and dividends, 29, 35, 49–53, 63, 65, 104; wire leases, 86, 91, 102, 115–16, 157, 163–65

Western Union Telegraph Co. v. Pennsylvania Railroad (1904), 40

Wheatstone, Charles, 42

White, Roy, 167–68

White, Samuel S., 149

Wickersham, George, 134

Wiley, G. L., 109–10

Wilkeson, Samuel, 88–89

Willever, John Calvin, 161

Wilson, Charles H., 163–64

Wilson, Edmund, 97

Wilson, J. J. S., 12, 19–20

Wilson, William L., 67

Wilson, Woodrow, 34, 67–69, 163

Wiman, Erastus, 38

Wodehouse, P. G., 80

Woodbury, Levi, 1–2

Wright, Abner, 126–27, 132

Yates, Richard, 6, 8